A NATURALIST'S GUIDE TO THE
WILD FLOWERS
OF BRITAIN
& NORTHERN EUROPE

A NATURALIST'S GUIDE TO THE

WILD FLOWERS
OF BRITAIN
& NORTHERN EUROPE

Andrew Cleave & Paul Sterry

JOHN BEAUFOY PUBLISHING

Reprinted in 2022

This edition first published in the United Kingdom in 2019 by
John Beaufoy Publishing Limited
11 Blenheim Court, 316 Woodstock Road, Oxford OX2 7NS, England
www.johnbeaufoy.com

10 9 8 7 6 5 4 3

ISBN 978-1-912081-14-1

Front cover photographs: *Main image:* Primrose; *bottom left:* Common
Poppy; *bottom centre:* Bluebell; *bottom right:* Yellow Iris.
Back cover photograph: Sea Holly.
Title page photograph: Marsh Marigold.
Contents page photograph: Purple Saxifrage.

Edited, designed and typeset by D & N Publishing, Baydon, Wiltshire, UK

Printed and bound in Malaysia by Times Offset (M) Sdn. Bhd.

·CONTENTS·

ACKNOWLEDGEMENTS

Thanks are due to the many individuals and organisations who have helped in various ways with the preparation of this book. Special thanks go to Paul Sterry of Nature Photographers for advice on the picture selection, and to Bill Helyar for help with the location of numerous plant specimens. Thanks also to James and Åsa Sutherland for an introduction to the flora of southern Sweden.

THE AREA COVERED BY THIS BOOK

This book describes the commoner or more conspicuous and eye-catching wild flowers most likely to be found in Britain and Ireland. It also covers the neighbouring regions of mainland Europe, including France (north of the Loire Valley), Belgium, Holland, northwest Germany, Denmark, Sweden and Norway.

Within this region there is a huge diversity of habitats, soil types and climate, and as a result there is a very rich and varied flora. Despite the fact that parts of the region are very densely populated and industrialised, there are still large tracts of natural and semi-natural vegetation in which wild flowers can flourish. Even the centres of large cities provide niches in which plants can grow, and many species are quick to colonise newly created habitats such as roadside verges and industrial waste tips. Gardeners will know how quickly wild plants can take over once a garden plot is neglected. Wherever the botanist lives, places to search for interesting wild flowers will never be far away.

In the far southwest of Britain and Ireland, on the Channel coast of France and, to a lesser extent, up the west coast of Britain, the influence of the Gulf Stream is at its greatest, and the climate here is relatively mild with high levels of rainfall; winter temperatures rarely drop below freezing and the summers usually have plenty of rain. This is described as an oceanic climate, and plants that are more at home further southwest in Europe can survive here, including many species that like a high humidity and are intolerant of frost. Moving further inland, and as a result losing some of the influence of the sea, the climate becomes more extreme, with warmer and drier summers, and far colder winters. Plants that can tolerate longer periods of drought are able to cope with this type of climate. Moving north within the region leads into a zone where the summers are shorter and the winters longer and more severe. Here, the plants need to be able to cope with lower levels of light, longer periods of cold weather, and perhaps spells beneath a blanket of snow.

HABITATS

The region covered by this guide includes a very great variety of habitat types, each with its own characteristic flora. A few wild flowers are very widespread and can be found over a large range and in many different habitats, often as a result of man's activities, but most are quite specialised and restricted to a particular habitat or set of conditions. The attractive Ivy-leaved Toadflax, for example, was introduced to Britain from the mountains of southern Europe as a garden plant and, like many garden plants, escaped into the wild. It is now firmly established and very widespread in Britain, finding man-made habitats such as walls and bridges much to its liking. Alpine plant enthusiasts, on the other hand, try to grow plants that are more at home on high mountains in specially created alpine beds in their gardens; these plants are far less likely to escape and establish themselves in the wild in lowland areas as the normal conditions found here would not suit them.

Altitude has a great bearing on which plants can survive in a given habitat. Conditions on high mountains can be very severe, with low temperatures, high rainfall, poor soils, strong winds and low sunshine levels. Further north in the region, these conditions can be found at lower altitudes, and so species that are considered to be alpine plants in the south may be found growing near sea-level in northern regions.

Soil types determine which plants can be found in a given area. Some plants are known as calcifuges, meaning that they avoid soils containing lime or calcium salts. These plants are normally found in boggy or peat-rich habitats, or growing in areas where there are hard rocks such as granite. Plants that prefer a lime-rich soil are known as calcicoles and are found growing in areas of chalk and limestone rocks, and sometimes on man-made sites.

COASTAL HABITATS Plants living close to the sea have to cope with very difficult environmental conditions. Exposure to salt spray can severely dehydrate tender plants, and strong winds, wind-blown sand and occasional inundation by the sea can damage them further. Many specialised plants can be found here, however, and they usually show distinct adaptations to this environment. Most will have fleshy leaves, which retain water, or leaves with a waxy surface or inrolled margins, which prevent water loss. Deep tap roots will help reach water far down in sand or shingle, and a rosette-forming habit will reduce the surface area exposed to the wind. A further important adaptation is the production of seeds that can withstand exposure to salt water; these can often be dispersed over great distances if they are washed into the sea, carried along on ocean currents and then drift ashore elsewhere.

Sea Lavender flowers in profusion on saltmarshes in summer.

Sand-dunes and saltmarshes support highly specialised plants that can cope with the ever-varying conditions. Shifting sand can smother small plants, but some species thrive in these conditions and often have deep, fibrous root systems to anchor them and help find water in the very free-draining substrate. The attractive Sea Holly is one of the specialist species that can live here, and Sea Bindweed is a colourful resident of mobile dunes. The regular inundation of the tide in saltmarshes means that plants growing here have to cope with frequently changing conditions – they may be partly submerged in salt water, or exposed to the air in thick mud. Common Sea Lavender is one of the few species well suited to this environment, and often grows in great abundance.

Bluebells flower early in the season in coppiced woodlands, and can often be the dominant species, indicating an ancient woodland site.

WOODS AND FORESTS Woodlands provide sheltered conditions for smaller herbaceous plants and often have a rich and diverse flora. The woodlands themselves are very varied, ranging from gloomy pine forests in which little light reaches the ground, to very open, light deciduous woodland in which there is leaf cover for only part of the year. Some deciduous woods, such as Beech woods, cast a very dense shade in the summer, so their flora is quite different from that of birch woods, for example. One feature of the flora of deciduous woodlands is the glorious display of spring flowers many of them support. These plants try to complete their flowering and much of their growth in the spring and early summer before the tree canopy develops and excludes much of the sunlight. Spectacular displays of Bluebells, Wild Daffodils and Wood Anemones can often be seen in these woodlands in springtime; later in the summer, when the leaf canopy is at its thickest, there will be far less to see on the woodland floor.

WETLANDS Watery habitats can vary greatly, from those that are completely aquatic, such as lakes and rivers, to fens, marshes and bogs. Some plants are specialised to live underwater for most of their lives, but almost all aquatic flowering plants will raise their flowers above the water to facilitate pollination. Some plants prefer to be rooted in water or wet ground but have their leaves and flowers raised above the surface. A few species can cope with fast-running water and have thread-like leaves to offer least resistance to the current, while others can survive only in still water; the floating leaves of water-lilies, for example, can tolerate only the most sheltered of conditions. The chemical composition of the water is also important and will determine what plants can be found in wetlands. Very acidic conditions are ideal for species like Round-leaved Sundew and Bog Asphodel, whereas slightly more alkaline conditions suit water-lilies and some of the marsh-orchids.

ABOVE: *The margins of cultivated land often provide safe habitats for wild flowers, and a great variety of species can be found here.*
LEFT: *Deep-rooted water lilies flourish in still waters, especially where there is little disturbance from boats.*

CULTIVATED LAND By definition, cultivated land will not support much wild flora. However, even on the most intensively farmed land there will be places where wild flowers – especially fast-growing annuals – can grow, including field margins, hedgerows, tracks and gateways. Some plants traditionally grew among the crops, but modern agriculture has found ways of removing these. The brightly coloured Corn Marigold, which as its name suggests grew in corn fields, still survives in a few areas of traditional farming practice, but once would have been extremely common and familiar.

URBAN SITES AND WASTELAND Wild flowers can be found right in the heart of towns and cities, and on abandoned and waste ground. Many will be annual plants that can complete their life cycles rapidly and set seed before conditions change too much. A large proportion of them will have wind-dispersed seeds, which enable them to reach these sites easily; Rosebay Willowherb and many thistles, for example, produce large quantities of airborne seeds. Waste and disturbed ground can provide good opportunities for colonisation by annuals that cannot compete easily with more vigorous, established vegetation elsewhere.

Urban sites are often free from agricultural sprays and provide a surprisingly safe refuge for species usually found in the countryside.

UPLANDS Mountainous regions can be very rewarding places for botanists. By their very nature, such areas have suffered less from the effects of agriculture or urbanisation, and the plants that grow there show many interesting adaptations to the difficult conditions in which they live. Some of the harshest climatic conditions of the region occur at high altitudes, and the further north the mountains occur, the more harsh the conditions become. Although agriculture may not be a problem in uplands, there are often high levels of grazing from sheep or wild deer, so some plants can survive only on inaccessible ledges. Commercial forestry may be a problem in some areas when non-native conifers are planted on what were once open hillsides. Tourism can also

Upland sites may appear rather barren at first, but they are often home to a good range of small but well-adapted species that can survive in harsh conditions.

cause problems, as winter skiing and summer hillwalking and climbing bring more people into remote areas, increasing trampling and erosion. However, the upland regions do still offer some very challenging and exciting experiences for botanists. Upland plants often have very bright and attractive flowers, and they usually grow in a compact form to help them survive the harsh conditions. They are also often very slow-growing and long-lived.

FLOWER STRUCTURE

There is a huge variation in flower size, shape, colour and composition, but there are some structures that all flowers bear, even though they may differ dramatically from one species to another. Flowers are the reproductive organs of a plant and contain the male and female reproductive cells. The male part of the flower is the stamen, which consists of the filament, a thread-like stalk, and the anthers, which produce the pollen, containing the male sex cells. The stamens are often arranged in a ring around the central, female part

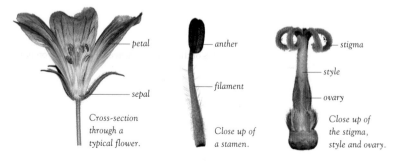

petal

sepal

Cross-section through a typical flower.

anther

filament

Close up of a stamen.

stigma

style

ovary

Close up of the stigma, style and ovary.

of the flower, which consists of the ovary, in which the seeds will develop, and the style, a stalk that supports the stigma on which pollen settles. Most flowers contain both male and female organs, but some, such as the Stinging Nettle, have flowers that consist of only male or female parts, with separate sexes on different plants.

The reproductive organs are usually surrounded by the petals and sepals, which are normally the most conspicuous parts of the flower. The sepals are the outer layer, and these are usually smaller and less brightly coloured. They close over the rest of the flower when it is still an unopened bud. The petals, which are usually brightly coloured, give the flower its form and character; their primary function is to attract pollinating insects, especially bees and butterflies. The petals are usually coloured in such a way that insects can easily spot them among green foliage. White flowers stand out well, as do bright yellow flowers, but colours in the blue and purple part of the spectrum are also important in attracting insects as they are easily detected by the compound eyes of bees and butterflies. In addition to their bright colours, many flowers will produce a pleasant scent, which will attract insects from a great distance and also work at night when the flowers cannot be seen. The petals may have nectaries at their base to provide a drop of sugary liquid as a 'reward' for the visiting insect; while collecting this important energy source the insect will be dusted with pollen, which will then be transferred to the next flower it visits. In this way cross-pollination occurs and the flower can then begin seed production. The pollinating insects themselves also show special adaptations in the form of long tongues to reach inside the flower, or hairy bodies to carry the pollen.

Whilst searching for nectar the bumble bee will pick up a dusting of pollen grains on the hairy surface of its body; these will then be carried to the next flower it visits.

Each seed on the dandelion flower head is topped with a tuft of hairs, which help it float away in wind currents. Whilst still attached they form the familiar 'dandelion clock'.

IDENTIFICATION

The identification of wild plants is based on variations in their structure, especially key features of the flowers and the arrangements of the stems and leaves. The glossary on p. 13 describes some of the more technical botanical terms used in the text, which will help with the correct identification of unfamiliar species. Important differences between species, and certain key features for identification, have been italicised in the descriptions.

In order to identify a plant successfully it will be necessary to look closely at the flowers and count the number and arrangement of petals and sepals. All members of the cabbage family, for example, have four petals arranged in the shape of a cross, hence their scientific family name of Cruciferae. All orchids have three petals and three sepals, while all members of the lily and iris families have floral parts in multiples of three. The arrangements of the internal parts of the flower are also important, and the number and arrangement of the stamens and the shape of the stigma in particular should be checked.

The arrangement of the individual flowers is another very good guide to identification as it varies greatly from species to species. Some plants bear solitary flowers on a single stem, such as buttercups, while others, like the Foxglove, may have many flowers on an upright spike. The spikes themselves may be one-sided and drooping, as in Bluebells, or have flowers all around them, as in the Common Broomrape. Mulleins often have a branched spike, while umbellifers have many small flowers on an umbrella-like arrangement of short stalks. Members of the daisy family have complex flowers comprising numerous tiny individual flowers, or florets, grouped together to resemble a single large flower.

The form of the plant itself may also be a helpful aid to identification, such as the presence of creeping stems, rosettes of leaves at the base or clasping leaves up the stem. In the species headings, the maximum height or growth form is given as another identification aid. Some plants have a very distinctive scent, so this should also be checked. It may be a pleasant scent produced by the flower to attract pollinating insects, or a very unpleasant smell released from the crushed foliage to act as a deterrent to grazing animals.

CHECKLIST FOR IDENTIFICATION
1. Count the number of petals and sepals.
2. Look at the arrangement of petals and sepals to see if they are joined at the base, or are of unequal size and shape.
3. Count the anthers and styles inside the flower.
4. Look for the arrangement of flowers on the stem: are they solitary, in a spike, in the leaf axils or at the end of the stem?
5. Is the plant smooth or hairy?
6. What is the arrangement of the leaves? Are they grouped around the base of the plant, opposite or alternate, or arranged up the stem?
7. Look at the shape of the leaves.
8. Is the plant annual or perennial?
9. How tall is the plant and does it branch?
10. Does the plant or the flower have a scent?
11. Check on the fruits or seeds.

EQUIPMENT TO AID IDENTIFICATION
The essential item of equipment for the field botanist, apart from a guide to wild flowers, is a hand lens that magnifies to at least 10x. A jeweller's loupe is ideal as it can be folded

A good hand lens is an essential item of equipment for the field botanist.

away and carried on a string around the neck. This type of lens is used by holding it very close to the eye and then bringing the specimen up close, usually to within just a few centimetres, until it comes into focus. Many useful features can then be seen, such as the degree of hairiness of the plant or the presence of tiny points on the leaf tips.

A notebook for jotting down descriptions and sketches is another very helpful item, but a digital camera is a good modern alternative. This eliminates the need to pick specimens and take them home for identification; the photographs can be used instead, provided they have recorded the essential identification features. Traditionally, botanists used to carry a large vessel called a vasculum, into which specimens were placed so that they could be examined at home later on. Often the specimens were then dried and pressed to provide a reference collection, each labelled with the date and place of collection and identification details. Today, the equivalent is a good collection of photographs, properly dated and labelled, and perhaps accompanied by notes in a diary. Such a record can be checked against one of the many websites devoted to wild flowers and their habitats, and will be a source of valuable information in later years and, possibly, the basis of further study.

WILDFLOWER NAMES

All plants have a scientific name that can be understood by botanists in any country. They also have common or vernacular names, in some cases several of them, which are used in only one country or perhaps just in a small region. In this guide, the common name is given first, in bold print, followed by the scientific name in italics. The scientific names are always in two parts, the first referring to the genus, or group, to which the plant belongs, and the second to its species. No two plant species can have the same name. Thus, Lady's Bedstraw is *Galium verum*, but Hedge Bedstraw is *Galium mollugo*. In Sweden, the vernacular names for these species are Gulmåra and Buskmåra, respectively, but their scientific names are exactly the same as in England.

It was the Swedish naturalist Carl von Linné, or Linnaeus as he is known today, who devised the system of giving all living things two names, known as binomial nomenclature. This system is still in place today, as are many of the names Linnaeus devised. The language used is a mixture of mostly Latin and some Greek, and the names usually convey some information about the plant. The scientific names used in this guide follow those in Clive Stace's *New Flora of the British Isles*, 4th edition (2019). The scientific names of species do sometimes change as taxonomists discover more about the relationships between different plants. The modern techniques of DNA analysis have enabled scientists to establish very accurately which plants are closely related and hence assign them to the correct group more easily.

GLOSSARY OF BOTANICAL AND HABITAT TERMS

Aggregate A group of very closely related species, which for the purposes of this book are treated as one. Abbreviated as agg.

Annual A plant that completes its entire life cycle within one year, i.e. germinating from its seed, producing flowers and then fruits, and dying off within 12 months.

Anther A small sac, often yellow in colour and growing in pairs, which contains the pollen. Anthers form at the top of the stamen.

Axil The inner angle between a leaf and a stem.

Base-rich Referring to soils that are rich in alkaline nutrients such as calcium salts.

Berry A fruit containing one or more seeds.

Biennial A plant that lives for two seasons, usually forming leaves in the first year and then flowering and fruiting in the second year, after which it dies.

Bog An area of waterlogged acidic peat and mosses (cf. 'Fen').

Bract A small leaf with a bud in its axil, or a collection of small leaves below an umbel.

Bulb An underground structure made up of swollen, fleshy leaf bases, as in a Wild Daffodil.

Calyx The whorl of sepals in a flower.

Capsule A dry fruit, often divided internally in sections.

Cladode A leaf-like structure that is really a stem.

Composite

Composite A member of the daisy family (Asteraceae).

Compound leaf A leaf that is divided into separate leaflets.

Corolla The petals of a flower.

Disc floret One of the inner, tubular florets of flowers of the daisy family.

Fen A very wet area with vegetation growing on alkaline-based peat, not acid-loving mosses (cf. 'Bog').

Compound leaf

Filament The thread-like stalk of the stamen, supporting the anthers.

Floret A small flower in a head made up of many similar flowers, such as in the daisy family.

Fruit The dry or fleshy structure surrounding a plant's seeds.

Lanceolate leaf A long, narrow leaf gradually tapering towards the tip.

Leaflets The separate leaf blades of a compound or pinnate leaf.

Lip The lower, and sometimes upper, lobes of the corolla of a flower such as an orchid.

Lanceolate leaf

Marsh An area of wet ground on a mineral-based soil, rather than peat.

Oblong leaf A leaf with parallel sides that is usually about three times longer than it is wide.

Palmate leaf A leaf with three or more leaflets that are arranged like digits.

Petal The coloured inner parts of the flower surrounding the stamens and stigma.

Pollen Microscopic yellow granules containing the male sex cells, produced inside the anthers.

Ray floret One of the outer florets of a composite flower, resembling a single petal but actually consisting of fused petals. These occur on the outside of flowers like daisies, but all over the head of flowers such as the Common Dandelion.

Recurved Arched backwards or downwards in a curve.

Rhizome A creeping underground stem.

Runner A stem that creeps above the ground.

Saltmarsh A marsh that is regularly inundated by sea water and usually found in very sheltered sites such as the upper reaches of estuaries.

Sepal One of the outer parts of the flower, usually green and closed over the bud before it opens.

Spike An unbranched flower stalk.

Stamen One of the male reproductive organs of a flower, consisting of a filament and a pair of anthers.

Stigma The surface of the female part of the flower, which receives the pollen.

Toothed leaf Referring to the edge of leaf which has regular, but usually rather small, indentations around the entire margin.

Toothed leaf

Trifoliate leaf A compound leaf with only three leaflets.

Umbel A usually flat-topped cluster of flowers whose individual flower stalks normally arise from the same point on the stem.

Umbel

Vein One of the prominent strands within a leaf containing the vessels that carry water.

Waste ground Disturbed but unused ground, such as on building sites, near new roads or in quarries and mines.

CONSERVATION OF WILD PLANTS

In most countries it is illegal to dig up wild flowers without the landowner's permission, and in the case of many rarities they are fully protected wherever they occur. It is not necessary to uproot any plant in order to identify it, and this practice should be strongly discouraged. It may be necessary, however, to collect the odd leaf or perhaps a single flower or floret in order to examine it more closely. This should be done with discretion, and if a plant appears to be very scarce it should be left well alone. Aquatic plants may sometimes need to be removed from the water in order to examine them properly, but they should be returned as quickly as possibly. It is also important to respect all of the plants in a habitat and not trample down surrounding vegetation in order to get a closer look at one specimen. Wherever possible, take the book to the plant instead of taking the plant home to the book. If you don't have a guide to wild flowers with you, taking photographs rather than specimens is a far better way of grappling with the identification of a tricky species.

CONSERVATION ORGANISATIONS There are now many local and national organisations dedicated to the conservation of wild plants and their habitats, and botanists – both amateur and professional – are strongly urged to join one. Doing so not only helps the wild plants, but also provides members the opportunity to join in on field visits, attend lectures or send photographs for identification. Membership may also provide access to nature reserves and collections not available to the general public. All such organisations publish journals or newsletters and have websites, which allow up-to-date information to be made available to their members. (*See* p. 156 for useful addresses and contacts.)

Common Bistort
■ *Bistorta officinalis* Up to 60cm

DESCRIPTION A conspicuous perennial plant with frothy, dense spikes, up to 20cm long, of small pink flowers, held upright above the plant on slender flower stalks. The long, triangular, alternate leaves are almost hairless; the lower leaves have long stalks and the upper ones clasp the stem. Individual plants are unbranched, but large, dense patches of Common Bistort may develop in suitable habitats, excluding all other species.
FLOWERING PERIOD Jun–Aug.
HABITAT Damp places, especially low-lying meadows, banks of ponds and rivers, and woodland rides, usually avoiding lime-rich soils.
FREQUENCY Widespread and locally common.

Common Sorrel
■ *Rumex acetosa* Up to 60cm

DESCRIPTION A rather variable and sometimes overlooked perennial plant of grassy habitats. The leaves are long and narrow with arrow-shaped bases; the upper leaves clasp the stem and the lobes of the lower ones point backwards; all have a distinct acidic taste. The small greenish flowers are borne on loose spikes and soon turn red, as does the whole plant at times. The fruits are in the form of small nutlets surrounded by a papery bract.
FLOWERING PERIOD May–Jul.
HABITAT Bare, open grassy places, almost always on acid soils, and usually avoiding very waterlogged areas.
FREQUENCY Widespread and common in suitable habitats, and very common in coastal areas.

Broad-leaved Dock

■ *Rumex obtusifolius* Up to 1m

DESCRIPTION A tall, familiar plant of fields and gardens. The broad, tapered leaves are up to 25cm long and have hairy veins on the underside. The largest, lower leaves have heart-shaped bases. The greenish-red flowers are borne in tall, loose spikes and the fruits have characteristic toothed margins and a single swollen tubercle. A deep tap root makes this a very persistent garden weed. The leaves are often perforated all over owing to attacks by capsid bugs.
FLOWERING PERIOD Jun–Aug.
HABITAT Field margins, bare and disturbed ground, open, sunny grasslands and urban wasteland.
FREQUENCY Widespread and sometimes common.

Fat Hen

■ *Chenopodium album* Up to 1m

DESCRIPTION A tall, slightly spreading, branched annual plant with a distinctly mealy appearance, although all parts are deep green underneath the dusty surface layer; the stem, however, may have a reddish tinge. The alternate leaves range in shape from elongate-oval to diamond-shaped, and the lower ones have slightly toothed margins. The small whitish flowers, which have no petals, are borne in tall spikes in the axils of the upper leaves, and the tiny rounded fruits are surrounded by 5 sepals.
FLOWERING PERIOD Jun–Oct.
HABITAT Disturbed arable land, farmyards and allotments, mainly in drier lowland areas.
FREQUENCY Widespread and common.

Red Goosefoot
■ *Oxybasis rubra* Up to 60cm

DESCRIPTION A variable, hairless, upright annual plant, which is often red-tinged, especially later in the season. The diamond-shaped alternate leaves have roughly toothed margins and slender stalks. A superficial resemblance to a webbed foot explains the English name. The very small, inconspicuous flowers are borne in short, dense spikes in the leaf axils, and the small, rounded fruits are enclosed by sepals.
FLOWERING PERIOD Jul–Oct.
HABITAT Disturbed but fertile soils, manure heaps, neglected gardens, often near the sea. Scarce in the N and in upland areas.
FREQUENCY Common and widespread in suitable habitats.

Sea Beet ■ *Beta vulgaris* ssp. *maritima* Up to 1m

DESCRIPTION A scrambling, untidy-looking perennial with shiny, rather leathery, dark green leaves that have reddish stems. The small green flowers have no petals and are borne

in slender, dense spikes. The small, spiked fruits dry out to become yellowish and may form small clumps. Colonies of Sea Beet can form a very dense mat of vegetation in some coastal habitats. A tough tap root helps locate and store water in the dry habitats in which this plant often grows.
FLOWERING PERIOD Jul–Sep.
HABITAT Coastal shingle, cliffs, margins of saltmarshes, stable dunes.
FREQUENCY Common and widespread along most coasts, although more frequent in the S.

Glasswort

■ *Salicornia europaea* agg. Up to 30cm

DESCRIPTION A fleshy yellowish-green annual with a jointed appearance; it may be branched but often grows as single succulent spikes. The leaves are scale-like, in opposite pairs, and fused to envelop the stem. Minute petal-less flowers appear in groups of 3 at the joints in the stem; the anthers are yellow, contrasting with the greenish stems, which turn red later in the season. The minute seeds are salt-tolerant.
FLOWERING PERIOD Aug–Sep.
HABITAT Entirely coastal, found in muddy, sheltered saltmarshes, sometimes forming a complete sward.
FREQUENCY May be abundant in suitable coastal habitats.

Sea Sandwort ■ *Honckenya peploides* Creeping

DESCRIPTION A creeping yellowish-green perennial with fleshy stems and leaves. The flowers are about 6–8mm across and have 5 white petals and slightly longer green sepals. Seeds are borne in yellowish-green capsules resembling small peas. In some sheltered areas where there is no competition and no trampling or disturbance, Sea Sandwort can form large mats of vegetation just above the high-tide line.
FLOWERING PERIOD May–Aug.
HABITAT Entirely coastal, preferring the upper reaches of sand and shingle beaches, often growing very close to the high-tide line.
FREQUENCY Locally common on suitable beaches.

Greater Stitchwort

■ *Stellaria holostea* Up to 50cm

DESCRIPTION A straggling perennial with squarish stems and narrow, pointed, unstalked leaves. The flowers are up to 30mm across and the 5 white petals are deeply notched. Seeds are borne in round green capsules. When the flowers are not present, the whole plant has a rather grass-like appearance and is easily overlooked among other vegetation.
FLOWERING PERIOD Apr–Jun.
HABITAT Woodland rides, glades, hedgerows and shaded grassy areas, usually on heavier soils.
FREQUENCY Widespread and sometimes very common in suitable habitats.

Common Mouse-ear

■ *Cerastium fontanum* Up to 30cm

DESCRIPTION A rather variable, short, hairy perennial with many leafy non-flowering shoots. The small white flowers, up to 7mm across, have 5 deeply notched petals and sepals of equal length. Leaf-like bracts occur just below the flowers. The seeds develop inside small greenish capsules. When not in flower, this is a very inconspicuous creeping plant, easily overlooked.
FLOWERING PERIOD Apr–Oct.
HABITAT Grassland and bare, open ground, in both rural and urban areas.
FREQUENCY Common and widespread.

Corn Spurrey

■ *Spergula arvensis* Up to 30cm

DESCRIPTION A straggling, stickily hairy annual weed, with whorls of narrow, blunt-tipped leaves. Each leaf is grooved on the underside and has a small leaf-like stipule at the base. The small white flowers, borne in loose umbels, are up to 8mm across and have undivided white petals. There are 5 yellow styles, and the seeds are produced in long capsules. FLOWERING PERIOD May–Aug. HABITAT Arable land and dry, disturbed ground, usually avoiding lime and only in lowland areas. FREQUENCY Widespread in farming areas, but probably declining through the use of herbicides.

Rock Sea-spurrey

■ *Spergularia rupicola* Up to 20cm

DESCRIPTION A short, slightly clump-forming perennial with stickily hairy leaves and stems. The narrow leaves form whorls around the stem, which may develop a purplish tinge. The attractive pinkish-purple flowers are about 8mm across and have 5 petals and 5 sepals of equal length. In very sunny situations the flowers can be produced prolifically. The seeds are borne in small greenish capsules. FLOWERING PERIOD Jun–Sep. HABITAT Sunny rock faces and cliffs, and sometimes on ruined coastal buildings, but always close to the sea. FREQUENCY Common in suitable maritime habitats in S and W areas.

Bladder Campion
■ *Silene vulgaris* Up to 80cm

DESCRIPTION An upright perennial with many flowering shoots and a woody base to the plant. The pointed, oval leaves are greyish green, hairless and often have a wavy edge. The slightly drooping flowers, borne in small clusters, have 5 deeply notched white petals and an inflated calyx, which forms a purple-veined bladder at the base of the flower. Seeds are produced in rounded capsules that dry out and become quite hard.
FLOWERING PERIOD Jun–Aug.
HABITAT Dry grassy places, usually arable land, and often on chalky soils.
FREQUENCY Widespread on suitable soils, but commonest in the S.

Moss Campion ■ *Silene acaulis* Cushion-forming

DESCRIPTION An attractive small, cushion-forming perennial plant with a woody base and dense masses of pink flowers. Individual flowers are up to 10mm across and have 5

slightly notched petals; seeds are produced in small capsules. When not in flower, the plant has a distinctly moss-like appearance with a tightly packed mass of narrow leaves. The flowers sometimes appear only on the S-facing or most sheltered side of the clump.
FLOWERING PERIOD Jun–Aug.
HABITAT Mountain tops and exposed rocky sites in upland areas, but also near sea-level in the far N.
FREQUENCY Locally common in suitable upland habitats in the N.

Red Campion
■ *Silene dioica* Up to 1m

DESCRIPTION An upright, hairy perennial plant with conspicuous, bright pink flowers up to 30mm across. The petals are notched, and separate male and female flowers are found on different plants, the male flowers being slightly smaller. Seeds are produced in flask-like capsules that have reflexed teeth. The oval, pointed leaves clasp the stem, but the lower ones have short stalks.
FLOWERING PERIOD Mar–Oct.
HABITAT Hedgerows and woodland rides, shady areas on rich soils.
FREQUENCY Widespread and common, sometimes dominating lanes and hedgerows in sheltered W areas.

White Campion
■ *Silene latifolia* Up to 1m

DESCRIPTION A tall, branching perennial with stickily hairy leaves and stems. The white flowers are up to 30mm across and have deeply notched petals. Male flowers, produced on a separate plant, are smaller than female flowers. The calyx tube is rather hairy with sticky glands on the surface and is marked with up to 20 darker vertical veins. Before opening, the flower buds are drooping in habit. Seeds are produced in flask-like pods that have upright teeth. This species sometimes hybridises with Red Campion (above), and a range of paler pink flower colours may result.
FLOWERING PERIOD May–Oct.
HABITAT Roadsides, hedgerows, disturbed ground and open grassy places.
FREQUENCY Widespread and common.

Ragged Robin
■ *Silene flos-cuculi* Up to 65cm

DESCRIPTION An upright perennial with slender, rough stems and lanceolate leaves, the upper ones in opposite pairs. The delicate pink flowers are up to 30mm across and the 5 petals are deeply and irregularly notched, giving them the characteristic 'ragged' appearance. Seeds are produced in slender capsules. When not in flower, the plant blends well with meadow grasses and is tricky to spot.
FLOWERING PERIOD May–Aug.
HABITAT Damp meadows and marshy areas.
FREQUENCY May be abundant and conspicuous in suitable habitats, but is declining owing to agricultural changes.

Winter Aconite
■ *Eranthis hyemalis* Up to 12cm

DESCRIPTION An attractive and conspicuous springtime flower. Its bright yellow flowers, made up of sepals, are up to 15mm across. Beneath the flowers is a frill of 3 notched leaves, but the palmately lobed true leaves appear at ground level after the flowers have died back. Seeds are produced in small, dry capsules. Dense patches of Winter Aconites may form in well-established colonies, often in association with Snowdrops (p. 143), and almost always as a result of deliberate introduction.
FLOWERING PERIOD Jan–Apr.
HABITAT Open woodlands and sheltered grassy banks, usually as a naturalised garden escape.
FREQUENCY Introduced and becoming widely naturalised.

Marsh Marigold

■ *Caltha palustris* Up to 25cm

DESCRIPTION A showy perennial plant with hollow stems. The shiny yellow flowers are up to 30mm across and consist of 5 sepals rather than petals. The large, hairless leaves are a rounded kidney shape and have a toothed margin, and those near the flowers may have a slightly marbled surface. Seeds are borne in dry capsules. Plants in shaded, sheltered habitats may be much larger than those in exposed upland areas.
FLOWERING PERIOD Mar–Jun.
HABITAT Damp woodland, marshy areas, pond and river margins.
FREQUENCY Widespread and common in suitable habitats, but probably declining.

Globeflower

■ *Trollius europaeus* Up to 60cm

DESCRIPTION A hairless, buttercup-like perennial plant. Its almost spherical yellow flowers are up to 40mm across and made up of about 10 yellow sepals, and are held erect on tall, upright stems. Seeds are produced in dry capsules. The palmate leaves are deeply divided and the upper ones clasp the stem. Globeflower often produces spectacular floral displays in traditional hay meadows in the N.
FLOWERING PERIOD May–Aug.
HABITAT Meadows and damp grassy areas, especially in uplands.
FREQUENCY Locally common in suitable habitats in N areas.

Meadow Buttercup
■ *Ranunculus acris* Up to 75cm

DESCRIPTION A familiar downy perennial plant with numerous shiny golden-yellow flowers borne on tall, unfurrowed stalks. The flowers are up to 25mm across and comprise 5 petals and 5 upright green sepals. Seeds are produced in numerous dry capsules with hooked tips. The leaves are deeply divided and comprise up to 7 lobes, with the end lobe unstalked, unlike other similar species.
FLOWERING PERIOD Apr–Oct.
HABITAT Damp grassland and meadows.
FREQUENCY Widespread and very common.

Bulbous Buttercup
■ *Ranunculus bulbosus* Up to 40cm

DESCRIPTION A short, very hairy perennial, superficially similar to Meadow Buttercup (above) but differs in having a markedly bulbous base to the stem. The flowers, supported on furrowed stems, are up to 30mm across and have 5 bright yellow petals and 5 reflexed greenish sepals. The seeds are borne in smooth, clustered capsules. The leaves comprise 3 stalked lobes.
FLOWERING PERIOD Mar–Jul.
HABITAT Dry grassland and meadows, including chalk downland and embankments.
FREQUENCY Widespread and often very common.

Lesser Spearwort
■ *Ranunculus flammula* Up to 50cm

DESCRIPTION An upright or creeping, hairless perennial, sometimes rooting at the stem nodes where it touches the ground. The leaves are spear-shaped, sometimes with short stalks, and the stems may become slightly reddened. The solitary flowers are up to 15mm across and grow on furrowed stems. Seeds form in dry, smooth capsules. Lesser Spearwort often climbs through other wetland vegetation and is very showy when in flower, although less obvious at other times. FLOWERING PERIOD Jun–Oct. HABITAT Damp meadows, streamsides and boggy areas, often on very acidic soils. FREQUENCY Widespread and locally common in suitable habitats.

Ivy-leaved Crowfoot
■ *Ranunculus hederaceus* Creeping

DESCRIPTION A mostly hairless annual or biennial plant with mats of slightly rounded, Ivy-shaped leaves. The creeping stems sometimes send down roots at the leaf nodes. The tiny white flowers are up to 6mm across and have 5 slender white petals and 5 green sepals of equal length. Seeds are borne in compact rounded heads. FLOWERING PERIOD May–Aug. HABITAT Bare muddy places, pond and river margins, and boggy ground, usually avoiding lime. FREQUENCY Widespread, but only locally common.

Wood Anemone
■ *Anemone nemorosa* Up to 30cm

DESCRIPTION A small, hairless perennial plant with solitary white, occasionally pink-tinged flowers consisting of petal-like sepals and borne on drooping stalks. The 3 deeply lobed leaves emerge at the same level from the stem, and after the flowers have finished more leaves may emerge at ground level. In very shady conditions, the plants may not produce flowers at all, but just send up the distinctive lobed basal leaves.
FLOWERING PERIOD Mar–May.
HABITAT Ancient woodlands, especially on heavy soils, and upland areas such as shaded cliffs and grasslands.
FREQUENCY Widespread and locally common, reaching quite high altitudes, but absent from the far N.

Common Meadow-rue
■ *Thalictrum flavum* Up to 1m

DESCRIPTION A tall, almost hairless perennial plant with dense terminal clusters of what appear to be yellow flowers, but are actually the very showy anthers; the tiny white petals fall soon after they open. These are followed by dry, papery fruits. The leaves are much divided into many small, toothed lobes, giving the plant a fern-like appearance.
FLOWERING PERIOD Jun–Aug.
HABITAT Damp, lush meadows, ditches, fens and wet areas on basic soils.
FREQUENCY Widespread in suitable habitats, and locally common in the S and E.

Common Fumitory

■ *Fumaria officinalis*
Up to 20cm, but often scrambling

DESCRIPTION A rather flimsy, weak, scrambling annual plant with greyish-green leaves that are much divided; all the lobes are flattened into a single plane. The complex purple-tipped pink flowers are up to 7mm long and consist of an elongated tube-like structure with a spur and 2 lips. A single seed forms inside a globular fruit. Many similar species occur and can be separated by the structure of the sepals and shape of the fruits. An ancient medicinal herb, once cultivated in monastic gardens.
FLOWERING PERIOD Apr–Oct.
HABITAT Gardens, well-drained, sunny arable areas and waste ground.
FREQUENCY Very widespread and common in some areas, but possibly declining owing to changing agricultural practices.

Common Poppy

■ *Papaver rhoeas* Up to 60cm

DESCRIPTION A conspicuous, roughly hairy annual plant, often occurring in very large numbers. The flowers can be up to 100mm across, and their overlapping papery petals are deep scarlet, usually with a dark base; the dark blue-black anthers contrast strongly with the petals. The seed pods are smooth and rounded, resembling small pepper pots.
FLOWERING PERIOD Jun–Aug.
HABITAT Arable fields, disturbed dry ground, wide roadside verges and wasteland.
FREQUENCY Widespread and common in many areas, but sometimes disappearing altogether when environmental conditions change.

Hedge Mustard
■ *Sisymbrium officinale* Up to 90cm

DESCRIPTION A tall, usually roughly hairy annual or, sometimes, biennial plant with a basal rosette of deeply divided leaves and more slender, entire leaves up the rather stiff flowering stem. Tiny yellow 4-petalled flowers, only about 3mm across, are borne in small terminal clusters on side shoots. Small cylindrical seed pods form, remaining upright and pressed close to the stem.
FLOWERING PERIOD May–Oct.
HABITAT Waste ground, disturbed soil, grassy embankments and arable land.
FREQUENCY Widespread and common.

Watercress
■ *Nasturtium officinale*
Up to 15cm

DESCRIPTION A short, creeping perennial plant with hairless, pinnately divided leaves and hollow stems that root at the leaf nodes. The leaves remain green through the winter. The 4-petalled white flowers are up to 6mm across and are produced in terminal clusters. The seeds form in 2 rows inside narrow pods about 18mm long. Extensively cultivated in specially created shallow beds irrigated by natural spring water.
FLOWERING PERIOD May–Oct.
HABITAT Usually grows in shallow flowing water or on very damp ground near streams, normally on lime-rich soils and areas of clean gravel.
FREQUENCY Widespread; commonest in the S.

Hairy Bittercress
■ *Cardamine hirsuta* Up to 30cm

DESCRIPTION A short annual plant with a basal rosette of pinnately divided leaves, each consisting of up to 7 pairs of leaflets, and a few smaller leaves up the hairless stem. The 4-petalled white flowers are up to 3mm across and sometimes lose their petals very quickly. The seed pods are long and held upright, extending above the small flowers. FLOWERING PERIOD Feb–Nov. HABITAT Damp, disturbed ground, neglected gardens, shady urban parks. FREQUENCY Widespread and common.

Cuckoo Flower
■ *Cardamine pratensis* Up to 50cm

DESCRIPTION A hairless perennial with a basal rosette of pinnately divided leaves that have rounded lobes; the stem leaves are much more finely divided. Usually single stems arise from the base, but the plants can occasionally be tufted in form. The 4-petalled flowers are up to 20mm across and can range in colour from white to pale lilac, but always with yellow anthers. The seed pods are slender and up to 40mm long. May also be called Lady's Smock. FLOWERING PERIOD May–Oct. HABITAT Damp grassy places and riverside meadows. FREQUENCY Widespread and locally common.

Shepherd's Purse

■ *Capsella bursa-pastoris* Up to 35cm

DESCRIPTION A very variable species, depending on its habitat, which may be annual or biennial, and showing varying degrees of hairiness. The leaves in the basal rosette are long and usually deeply toothed, but may have almost entire margins, while the upper leaves clasp the stem. The small white 4-petalled flowers, in terminal clusters, are only about 2–3mm across, and they give rise to the characteristic inverted triangular, notched seed pods. The flowering stems elongate as the seed pods develop.
FLOWERING PERIOD Jan–Dec.
HABITAT Arable fields, tracks, waste ground, and urban parks and gardens.
FREQUENCY Widespread and very common.

Common Scurvy-grass

■ *Cochlearia officinalis* Up to 50cm

DESCRIPTION A rather variable, hairless perennial or biennial plant with fleshy leaves. The lower heart-shaped leaves form a basal rosette, and the upper leaves, which are more pointed, clasp the stem. The 4-petalled white flowers are up to 10mm across and the seed pods are rounded, developing on short stalks.
FLOWERING PERIOD Apr–Oct.
HABITAT Coastal cliffs and drier areas of saltmarshes, sea walls, and also inland on mountain crags and screes. Now increasingly seen well inland along roadsides owing to the spreading of salt in winter.
FREQUENCY Widespread and locally common in suitable habitats.

Garlic Mustard

■ *Alliaria petiolata* Up to 1m

DESCRIPTION A tall, hairless, leafy biennial with a distinct scent of garlic emitted from crushed leaves. The large, long-stalked leaves are heart-shaped at the base and have toothed margins. The 4-petalled white flowers, up to 6mm across, grow in terminal clusters and are followed by long, upright seed pods.
FLOWERING PERIOD Apr–Jun.
HABITAT Hedgerows, grassy banks, roadside verges and woodland clearings.
FREQUENCY Widespread and common, especially in the S and E.

Sea Rocket

■ *Cakile maritima* Up to 25cm

DESCRIPTION A straggling, slightly fleshy, hairless annual with long, pinnate leaves that are greyish green in colour. The pink or white 4-petalled flowers are up to 15mm across and grow in terminal clusters; they have a delicate scent. Seeds develop inside long, waisted pods, the upper half of which is larger than the lower.
FLOWERING PERIOD Jun–Sep.
HABITAT Just above the high-tide line on sandy or shingle beaches.
FREQUENCY Very widespread and locally common around all coasts. Copious quantities of floating, salt-tolerant seeds enable this species to spread easily around coasts and colonise new habitats.

Sea Kale
▪ *Crambe maritima*
Up to 60cm

DESCRIPTION A large, clump-forming perennial plant, which can develop into dense, domed masses of thick leathery leaves and abundant flowers, growing from a woody root-stock. The fleshy leaves have a wavy margin and the lower ones grow on long stalks. The grey-green coloration may become tinged with purple later in the season. The 4-petalled flowers are white with yellow anthers and grow in dense, flat-topped heads that sometimes cover the entire plant. Seeds develop inside woody oval pods.
FLOWERING PERIOD Jun–Sep.
HABITAT Mainly on shingle beaches, but also on sand and sometimes at the base of cliffs; always close to the high-tide line.
FREQUENCY Locally common in suitable habitats around most of the region's coastline.

Weld ▪ *Reseda luteola* Up to 1.2m

DESCRIPTION An erect, unbranched, hairless biennial with a hollow stem. The leaves are narrow and untoothed, but have a wavy margin. In the 1st year they form a basal rosette, but in the 2nd year they grow up the flowering stem. The tiny yellow flowers, with 4 sepals and petals, are up to 5mm across, and are densely packed on a tall spike. Seeds develop inside globular pods.
FLOWERING PERIOD Jun–Aug.
HABITAT Disturbed ground and dry grassy places, usually on lime-rich soils.
FREQUENCY Widespread and common, but becoming more scarce in the N.

Wild Mignonette

■ *Reseda lutea* Up to 70cm

DESCRIPTION An erect, solid-stemmed biennial with long, wavy-edged, pinnately divided leaves covering the stem. The small yellowish-green flowers, up to 6mm across, have 6 petals and sepals and are delicately scented. They are arranged in a tall terminal spike. Seeds develop inside oblong pods, held upright close to the stem. FLOWERING PERIOD Jun–Sep. HABITAT Disturbed ground, neglected arable land and waysides, more often on lime-rich soils. FREQUENCY Widespread and locally common, but absent from the N.

Round-leaved Sundew

■ *Drosera rotundifolia* Up to 20cm

DESCRIPTION A curious and distinctive insectivorous perennial plant with a rosette of long-stalked rounded leaves up to 10mm across. The red leaves are covered with glistening, sticky hairs that both attract and trap small insects. The white flowers are about 5mm across and supported on long, drooping spikes that emerge from the centre of the rosette. Seeds are produced in tiny green capsules. FLOWERING PERIOD Jun–Aug. HABITAT Sphagnum bogs and very wet ground on heaths and moors, avoiding lime-rich areas. FREQUENCY Widespread and locally common in suitable habitats.

Navelwort

■ *Umbilicus rupestris* Up to 15cm

DESCRIPTION A distinctive hairless perennial plant that has rounded, fleshy leaves with a 'navel' in the centre. Its alternative common name of Wall Pennywort refers to the rounded shape of the leaves. Tall flower spikes support the tubular, drooping greenish-white flowers. The plants commonly grow on walls and rocks, so both the leaves and flower spikes are often held flat against the vertical rock face.
FLOWERING PERIOD Jun–Aug.
HABITAT Rocks and walls, often in the shade, and always avoiding lime-rich areas.
FREQUENCY Widespread and locally common in suitable habitats in W areas, but very scarce elsewhere and absent from the far N.

English Stonecrop

■ *Sedum anglicum*
Up to 5cm but usually spreading

DESCRIPTION A creeping, mat-forming perennial with fleshy, cylindrical leaves arranged along slender stems. The greenish leaves often turn red by the end of the summer. The showy, star-like, 5-petalled white flowers are up to 12mm across; the undersides of the petals are often tinged with pink. The flowers are held above the plant in a compact, slightly branched flower spike.
FLOWERING PERIOD Jun–Sep.
HABITAT Rocky ground, stabilised shingle, old walls and cliffs, usually avoiding lime-rich areas.
FREQUENCY Widespread and locally common in suitable habitats, most frequent in W areas.

Biting Stonecrop
■ *Sedum acre* Up to 10cm

DESCRIPTION A creeping, mat-forming perennial plant with yellowish-green succulent leaves. The common name refers to the peppery taste of the leaves when chewed. The star-like, 5-petalled yellow flowers are up to 12mm across and produced in profusion on short, erect shoots; the plant can often be spotted from a great distance because of the striking appearance of a mass of the flowers. FLOWERING PERIOD May–Jul. HABITAT Dry, open habitats, including sand-dunes and shingle, but sometimes also on industrial waste ground and dry roadsides. Often found in lime-rich areas. FREQUENCY Widespread and locally common, becoming less so in the far N.

Meadow Saxifrage
■ *Saxifraga granulata* Up to 45cm

DESCRIPTION A downy perennial plant with a basal rosette of stalked, kidney-shaped, toothed leaves that develop prominent bulbils at their bases late in the season. The 5-petalled white flowers, up to 30mm across, are held erect in a loose cluster on a tall shoot, which raises them above the surrounding meadow grasses. When not in flower, this is a very difficult plant to find among other vegetation. FLOWERING PERIOD Apr–Jun. HABITAT Grassy meadows and banks, usually on heavier basic soils. FREQUENCY Widespread and locally common in suitable lowland habitats, especially in E areas.

Purple Saxifrage
■ *Saxifraga oppositifolia* Creeping

DESCRIPTION A creeping, mat-forming perennial with numerous trailing stems and small, stalkless, opposite leaves, often encrusted with tiny granules of lime. The strikingly purple 5-petalled flowers, with yellow anthers, are up to 15mm across and lie close to the plant on very short stalks. This is often one of the first alpine flowers to come into bloom in mountainous regions.
FLOWERING PERIOD Mar–Aug depending on altitude and geographical location.
HABITAT Mountain rocks and screes, often in lime-rich areas and damp habitats, and close to sea-level in the far N.
FREQUENCY Locally common only in suitable upland habitats in the N and NW.

Opposite-leaved Golden Saxifrage
■ *Chrysosplenium oppositifolium*
Up to 15cm

DESCRIPTION A slightly hairy, low-growing, patch-forming perennial with square stems and rounded, bluntly toothed leaves growing in opposite pairs. The tiny greenish-yellow flowers are only about 4mm across and have no petals; they grow in flat-topped clusters with yellow bracts beneath them, these giving the impression of petals. A very similar species has alternate leaves and a distribution more to the E.
FLOWERING PERIOD Mar–Jul.
HABITAT Damp and partly shaded streamsides and gullies, damp flushes in woodlands and gorges.
FREQUENCY Locally common, occurring mainly in the N and W.

Grass of Parnassus

■ *Parnassia palustris* Up to 25cm

DESCRIPTION Not a grass at all, but a tufted, hairless perennial plant with large white flowers, up to 30mm across, on solitary tall stems, each one with a single clasping leaf. The flowers have 5 petals, marked with darker greenish veins, and yellow anthers. Seeds develop inside a dry capsule. The stalked, heart-shaped basal leaves are in a rosette and are smooth and toothless.
FLOWERING PERIOD Jun–Sep.
HABITAT Marshes, moors and damp grassland over peaty soils.
FREQUENCY Locally common in the N and NW, but scarce or absent in the S.

Agrimony

■ *Agrimonia eupatoria* Up to 50cm

DESCRIPTION A moderately tall, downy perennial plant with pinnately divided leaves, the individual lobes of which have a slightly ridged or grooved appearance; smaller leaflets are found between the larger lobes. The yellow flowers, which grow in tall, slender spikes, have 5 petals and are up to 8mm across. The fruits are dry and grooved, and have small hooks at the apex that readily attach to animal fur or clothing to aid dispersal.
FLOWERING PERIOD Jun–Aug.
HABITAT Roadsides, dry grassy places and waste ground, often on lime-rich soils.
FREQUENCY Widespread and common, especially in drier lowland areas.

Meadowsweet
■ *Filipendula ulmaria*
Up to 1.25m

DESCRIPTION A tall, hairless perennial plant with long-stalked, pinnately divided leaves comprising 2–5 pairs of leaflets with much smaller leaflets between them. Small leaf-like stipules with pale, downy undersides are also present. The strongly scented creamy-white flowers grow in dense, frothy clusters; individual flowers may have 5 or 6 petals and measure about 5mm across.
FLOWERING PERIOD Jun–Sep.
HABITAT Damp meadows, riversides, marshes and open woodland rides.
FREQUENCY Common and widespread in suitable habitats.

Dog Rose
■ *Rosa canina*
Climbing, up to 3m

DESCRIPTION A scrambling hedgerow shrub with long, arching stems bearing strong, curved thorns. The hairless leaves have 2 or 3 pairs of toothed leaflets. The beautifully fragrant flowers, usually growing in clusters of 4, have 5 pink or pink-tinged petals and a ring of yellow stamens. The fruits are in the form of shiny red rose hips, which lose their ring of dried sepals before they ripen.
FLOWERING PERIOD Jun–Jul.
HABITAT Hedgerows and waysides, scrub patches.
FREQUENCY Widespread and common.

Sweet Briar
■ *Rosa rubiginosa* Up to 3m

DESCRIPTION A dense hedgerow shrub bearing many strong, upright stems armed with short, curved thorns and finer bristles. The compound leaves are divided into 5–7 leaflets that are covered with sweet-smelling glandular hairs. The attractive, scented pink flowers are up to 3cm across and grow in clusters of 3. Seeds are produced in shiny red rose hips, which retain their dried sepals once they have ripened. FLOWERING PERIOD Jun–Jul. HABITAT Hedgerows and scrub, most commonly on lime-rich soils. FREQUENCY Locally common, mostly in drier areas of the S and E, but increasingly planted in parks and on roadsides.

Burnet Rose
■ *Rosa spinosissima* Up to 50cm

DESCRIPTION A low-growing flowering shrub with a tangle of suckering stems that bear straight thorns and numerous stiff bristles. The compound leaves are made up of 3–5 pairs of rounded leaflets. The cup-shaped, solitary white flowers, which may occasionally be tinged with pink, are up to 40mm across; in their centre are many yellow stamens. The hips are a deep purple colour, sometimes almost black, and they retain the dried sepals. FLOWERING PERIOD May–Jul. HABITAT Dry limestone pavements, coastal dunes and stabilised shingle. FREQUENCY Widespread but local, mainly near the coast.

Salad Burnet
■ *Poterium sanguisorba* Up to 35cm

DESCRIPTION A low-growing, mostly hairless perennial with greyish-green compound leaves divided into 4–12 pairs of similar-sized oval, toothed leaflets. There is a hint of cucumber scent and flavour in the crushed leaves. The flowers grow in compact globular heads, raised on long stalks; there are no petals, but the upper flowers in the head have red styles and the lower ones have yellow stamens. The sepals are small and green. Ridged, 4-sided fruits follow the flowers.
FLOWERING PERIOD May–Sep.
HABITAT Dry grassland and chalk downs, always in open, sunny positions.
FREQUENCY Widespread and locally abundant in suitable habitats, mainly in the S; very scarce in Scotland.

Lady's Mantle
■ *Alchemilla vulgaris* agg. Up to 30cm

DESCRIPTION A rather variable perennial plant with many similar species, including a familiar garden escape. The rounded, stalked leaves are lobed and have finely toothed margins; they may be densely hairy, but hairless forms are sometimes found. The tiny yellowish-green flowers are up to 5mm across and lack petals. They grow in loose clusters on long, branched flower stalks.
FLOWERING PERIOD May–Sep.
HABITAT A variety of grassland habitats, including damp meadows, woodland rides, grassy moorlands and lower mountain slopes.
FREQUENCY Widespread and locally common in suitable habitats, but more frequent in the N.

Mountain Avens
■ *Dryas octopetala* Up to 6cm

DESCRIPTION A low-growing, creeping perennial with attractive white flowers. The oblong, blunt-toothed, long-stalked leaves are dark green above and grey on the underside. The flowers are up to 20mm across and have 8 or more white petals and numerous bright yellow stamens. The flowers twist to follow the sun during the day in order to warm the pollen and attract bees. The small, dry seeds have long, feathery styles to aid wind dispersal.
FLOWERING PERIOD May–Jul.
HABITAT Exposed mountain rocks and grassland, but reaching down to sea-level in the extreme N. Most common on lime-rich substrates.
FREQUENCY Widely scattered and locally common only on high mountains in the N of the region, or nearer sea-level in the extreme N.

Wood Avens
■ *Geum urbanum* Up to 50cm

DESCRIPTION A downy, or sometimes hairy, perennial plant with compound leaves divided into 3–6 pairs of opposite leaflets and 1 much larger terminal leaflet. The leaves growing up the stem have smaller leaf-like stipules at their bases. The 5-petalled yellow flowers are up to 15mm across and are borne singly on slightly drooping stalks. The seed heads are in the form of prickly burs with numerous hooked red spines to aid dispersal by animals. The species may also be known by the alternative name of Herb Bennet.
FLOWERING PERIOD May–Sep.
HABITAT Hedgerows, woodlands (where it often appears on recently disturbed sites) and shady areas with damp, fertile soils. Mostly in lowland areas.
FREQUENCY Widespread and sometimes very common.

Water Avens
▪ *Geum rivale* Up to 50cm

DESCRIPTION A downy perennial plant with pinnately divided basal leaves and 3-lobed leaves further up the stem; these have tiny leaf-like stipules at their bases. The drooping, bell-shaped flowers have purplish-red sepals and pink petals; they can be up to 15mm across. Once pollinated, the fruits develop into a bur-like head with feathery, hooked styles. These are held upright, making it easier for them to get hooked into a passing animal's coat to aid dispersal.
FLOWERING PERIOD Apr–Sep.
HABITAT Damp, lush meadows and marshes; also on shaded, well-vegetated mountain ledges.
FREQUENCY Widespread and locally common, becoming more frequent in the N.

Marsh Cinquefoil
▪ *Comarum palustre* Up to 40cm

DESCRIPTION A hairless perennial plant with leaves that are pinnately divided into 3–5 elongate, toothed greyish-green leaflets. The star-shaped flowers are up to 30mm across and have 5 slender, deep purple petals and 5 much broader, spreading maroon sepals. They grow in loose clusters on long, slender flower stalks. Dry, papery seeds develop, enclosed by the sepals.
FLOWERING PERIOD May–Jul.
HABITAT Marshes and damp meadows, usually avoiding soils rich in lime, but tolerant of very waterlogged conditions.
FREQUENCY Widespread and locally common, mainly in lowland areas in the N and W, but also in the S and E in suitable habitats.

Tormentil

■ *Potentilla erecta* Up to 30cm

DESCRIPTION A creeping, patch-forming perennial plant with downy stems. The basal leaves have 3 rounded leaflets, but these do not persist on the plant; the stem leaves are unstalked, looking as if they are divided into 5 toothed leaflets but in fact are also trefoil, and have 2 toothed leaf-like stipules where they join the stem. The 4-petalled yellow flowers are arranged in loose clusters on weak stems, and are sometimes produced prolifically. The very similar Trailing Tormentil *P. anglica* may have 5 petals, and its trailing stems can form roots at the leaf nodes.
FLOWERING PERIOD May–Sep.

HABITAT Open grassland, heaths and moors, but not on lime-rich soils.
FREQUENCY Widespread and sometimes abundant, covering large areas of suitable habitats.

Silverweed

■ *Potentilla anserina* Creeping

DESCRIPTION A creeping perennial plant with distinctive pinnately lobed silvery leaves and numerous long, red-tinged runners that send down rootlets at the leaf nodes. The leaves are made up of as many as 12 pairs of opposite, toothed leaflets and are covered in silky silvery hairs. The bright yellow flowers are produced singly on long stalks and are up to 15mm across. There are 5 petals, which are double the length of the narrow green sepals.
FLOWERING PERIOD May–Aug.
HABITAT Damp grassy places and bare, open ground from sea-level to hilly areas. It is tolerant of brackish conditions.
FREQUENCY Widespread and very common.

Wild Strawberry
■ *Fragaria vesca* Up to 30cm

DESCRIPTION A low-growing, spreading perennial with long, rooting runners. The bright green 3-lobed leaves are made up of 3 toothed leaflets that have silky hairs on the underside. The 5-petalled white flowers have narrow sepals that are slightly longer than the petals, and are supported on short flower stalks in loose heads. The fruits are the familiar, but very small, bright red strawberries, covered on the outside with tiny seeds.
FLOWERING PERIOD Apr–Jul.
HABITAT Dry grassy places, open woodlands and embankments, preferring lime-rich soils in the N of its range.
FREQUENCY Widespread and common, except in the extreme N.

Broom ■ *Cytisus scoparius* Up to 2m

DESCRIPTION A much-branched, spineless deciduous shrub, which produces spectacular displays of yellow flowers in early summer. The green twigs that bear the flowers are ridged and 5-angled, and the leaves are divided into 3 lobes. The large yellow flowers are up to 2cm long and sometimes paired. Flat black seed pods develop, bursting open to release the seeds on hot summer days. A creeping, prostrate form of Broom is sometimes found on exposed cliffs and beaches on the coast.
FLOWERING PERIOD Apr–Jun.
HABITAT Heaths, embankments, cliffs, sandy soils and stony riversides, mostly in lowland areas.
FREQUENCY Widespread and common, apart from in the extreme NW and upland areas.

Common Gorse
■ *Ulex europaeus* Up to 2.5m

DESCRIPTION An extremely spiny
evergreen shrub with a stout woody base,
forming very dense, impenetrable thickets.
The shoots are green and the leaves are
trifoliate in young plants, but the twigs are
then covered in sharp, ridged green spines.
The yellow flowers are typical of the pea
family (Fabaceae), are produced in long,
stalked spikes, and give off an almond
or coconut scent on sunny days. Twisted
black seed pods are produced, which split
open to release the seeds with an audible
snap on hot afternoons.
FLOWERING PERIOD Jan–Dec, but most
prolific in Feb–May.
HABITAT A wide range of mostly lowland
habitats, especially on acid and sandy soils,
where it can dominate the landscape.
FREQUENCY Widespread and sometimes
abundant.

Dyer's Greenweed
■ *Genista tinctoria* Up to 1m

DESCRIPTION A spineless deciduous
shrub, with slightly hairy, lanceolate leaves
on green side shoots. In exposed sites the
plant may be prostrate, but it will also grow
through other low vegetation. The bright
yellow flowers are up to 15mm long and are
borne on leafy shoots. Seeds are produced
in flat, hairless pods. This plant was used
as the source of a dye for colouring fabrics,
hence its common name.
FLOWERING PERIOD Jun–Aug.
HABITAT Grasslands on heavy clay soils,
rough pastures, cliffs and roadside verges,
mostly in lowland areas.
FREQUENCY Locally common in the S,
especially England and Wales, and scattered
in S Scotland. Absent from Scandinavia.

Tufted Vetch

■ *Vicia cracca* Scrambling, up to 2m

DESCRIPTION A slightly downy, clambering perennial plant with pinnately divided leaves composed of about 12 pairs of leaflets; the leaves terminate in branched tendrils that aid climbing. The bluish-purple flowers are up to 12mm long and are arranged in dense, 1-sided spikes up to 8cm long. The seeds are produced inside long brown seed pods.
FLOWERING PERIOD Jun–Aug.
HABITAT Grassy places, hedgerows, woodland rides and scrub.
FREQUENCY Widespread and very common, but does not occur in very wet areas or at high altitudes.

Common Vetch

■ *Vicia sativa* agg. Up to 75cm

DESCRIPTION A scrambling, downy annual plant with compound leaves made up of 3–8 pairs of small oval or elongate leaflets, and terminating in tendrils. Stipules at the base of the leaf may have a dark central spot. The reddish-purple flowers are borne singly or in pairs and are up to 30mm long; they are followed by narrow, flattened pods that become black when ripe and may be slightly downy.
FLOWERING PERIOD Apr–Sep.
HABITAT Open, dry grassy areas, hedgerows and embankments, especially on sandy soils.
FREQUENCY Widespread and quite common in places, often as a survivor of cultivation where it was once grown as a fodder crop.

Meadow Vetchling
■ *Lathyrus pratensis* Up to 50cm

DESCRIPTION A scrambling, usually hairless perennial with angled stems. The leaves are made up of a pair of narrow leaflets and a terminal tendril; at the base of each leaf is an arrow-shaped leaf-like stipule. The yellow flowers are up to 18mm long and grow in long-stalked clusters of as many as 12 individual flowers. Seeds are produced in long pods that become black as they mature.
FLOWERING PERIOD May–Aug.
HABITAT Grassy places, including hedgerows and embankments, roadsides and old hay meadows.
FREQUENCY Widespread and very common, but scarce or absent in uplands.

Kidney Vetch
■ *Anthyllis vulneraria* Up to 30cm

DESCRIPTION A rather variable, low-growing perennial covered in silky hairs. Depending on the habitat, it may scramble through other low vegetation or grow flat on exposed rocks. The leaves are pinnately divided, with opposite pairs of narrow leaflets and a single larger terminal leaflet. The yellow flowers grow in compact heads up to 3cm across; these usually occur in pairs. Considerable variation in flower colour can be found, with reddish-purple forms occurring in some areas. Seeds develop inside short pods.
FLOWERING PERIOD Apr–Sep.
HABITAT Rocky outcrops and sunny, S-facing, dry slopes, especially on sea cliffs, chalk downs and inland embankments.
FREQUENCY Widespread and locally common in suitable habitats.

Common Restharrow
■ *Ononis repens* Up to 70cm

DESCRIPTION A tough perennial plant with a shrubby habit and woody-based, creeping, hairy stems. The dark green leaves are stickily hairy and made up of 3 oval leaflets. The pink flowers, which are up to 15mm long, are raised on leafy shoots, and these give rise to short, dark seed pods. The common form of this plant has no spines, but a rare spiny form does sometimes occur; the separate species **Spiny Restharrow** *O. spinosa* is a much taller plant.
FLOWERING PERIOD Jul–Sep.
HABITAT A wide range of open grassy habitats, including chalk downs, dune slacks and cliff tops, usually on free-draining, lime-rich soils.
FREQUENCY Widespread and locally common in the S, but scarce in much of Scotland and Ireland, and present only in coastal areas of Scandinavia.

Ribbed Melilot
■ *Melilotus officinalis* Up to 1.5m

DESCRIPTION A showy, hairless, upright biennial plant with numerous tall spikes, up to 7cm long, of bright yellow flowers. Seeds are produced in short, hairless brown pods. The leaves are fairly long-stalked and made up of 3 equal, toothed oblong leaflets. The whole plant can tower above other meadow vegetation, making it very conspicuous when in full flower.
FLOWERING PERIOD Jun–Sep.
HABITAT Grassy areas and waste ground, roadsides and embankments in lowland areas.
FREQUENCY Locally common and widely established in the S and E, where it occurs as an escape from agriculture, having formerly been used as a fodder crop.

Horseshoe Vetch

■ *Hippocrepis comosa* Up to 10cm

DESCRIPTION A spreading, hairless or slightly downy perennial plant arising from a woody root-stock, with pinnately divided leaves terminating in a single leaflet. The yellow flowers are up to 10mm long and arranged in a single plane in a loose circle of up to 12 individual flowers. Seeds are produced inside long, curly pods made up of several horseshoe-shaped segments.
FLOWERING PERIOD Jun–Sep.
HABITAT Open, sunny grasslands, mainly on chalk and limestone; increasingly found in roadside cuttings in chalk areas.
FREQUENCY Locally common in suitable habitats in the S and E, mainly in lowland areas.

Common Bird's-foot-trefoil

■ *Lotus corniculatus* Up to 10cm

DESCRIPTION A scrambling, usually hairless perennial plant with compound leaves made up of 5 oval leaflets, but appearing to be trifoliate because 1 pair of leaflets is at the base of the stalk. The flowers are orange in bud, but become yellow when fully open, and are borne in heads of 2–7 individual flowers. Seeds are produced in slender, straight pods in an arrangement resembling a bird's foot, hence its common name.
FLOWERING PERIOD May–Sep.
HABITAT A wide variety of grassland habitats from sea-level to high mountains, avoiding only the most acid or infertile soils.
FREQUENCY Widespread and, in places, very common.

Black Medick
■ *Medicago lupulina* Up to 20cm

DESCRIPTION A low-growing, usually downy annual or short-lived perennial plant with trefoil leaflets, each one bearing a minute point at the centre of the apex. As many as 50 of the tiny yellow flowers are arranged together in a compact, rounded head on a short stalk, each plant producing many such heads. Seeds develop inside small, curved black pods.
FLOWERING PERIOD Apr–Oct.
HABITAT A variety of open, dry grassland habitats, including roadsides, wastelands, embankments and other sites with infertile or well-drained soils.
FREQUENCY Widespread and locally common, but absent from the far NW.

Bird's-foot
■ *Ornithopus perpusillus* Up to 20cm

DESCRIPTION An insignificant trailing annual plant, usually downy, with pinnately divided leaves made up of as many as 13 pairs of oval leaflets and 1 terminal leaflet. The tiny flowers are up to 5mm long and creamy yellow with red veins, and are usually found in heads of 3–5 individuals. The long, beaded seed pods are arranged in a head that resembles a bird's foot.
FLOWERING PERIOD May–Aug.
HABITAT Open, sunny grassland sites, usually well grazed or trampled, and generally on sandy or acidic soils. Also in dunes and heathland tracks.
FREQUENCY Locally common in suitable habitats in the S of the region, but becoming scarce further N and absent from most of Ireland and Scandinavia.

Hop Trefoil

▪ *Trifolium campestre* Up to 25cm

DESCRIPTION A low-growing, hairy annual plant with trefoil leaves; the middle leaflet is long-stalked and all leaflets lack points at the tips. The tiny yellow flowers are individually about 4mm long, but are arranged in compact, globular heads up to 12mm across. Seeds develop inside tiny pods surrounded by the rounded heads of dried papery flowers, these resembling hops.
FLOWERING PERIOD May–Sep.
HABITAT Open, dry grasslands and embankments, spoil heaps, old walls.
FREQUENCY Widespread and locally common in the S and E, but becoming more scattered in the N and W.

Hare's-foot Clover

▪ *Trifolium arvense* Up to 25cm

DESCRIPTION A low-growing, softly hairy annual or biennial clover with trifoliate leaves, the individual leaflets being narrow and slightly toothed. The white or very pale pink flowers are tiny and arranged in stalked, fluffy, egg-shaped heads about 15–20mm high. Minute seed pods are hidden inside the calyx.
FLOWERING PERIOD Jun–Sep.
HABITAT Exposed dry, sunny sites on sandy or rocky ground, old walls and spoil tips. More common on acid rocks and soils, and usually not on upland sites.
FREQUENCY Widespread and locally common in the S and E, especially around the coasts, but becoming scarce in the far N and NW.

Red Clover
■ *Trifolium pratense* Up to 40cm

DESCRIPTION A familiar downy perennial herb with trefoil leaves, each narrow leaflet usually bearing a white crescent pattern. Triangular stipules tipped with bristles are found at the base of the leaves. The small pinkish-red flowers are borne in unstalked, globular heads, sometimes in pairs, with a pair of short-stalked leaves below the head.
FLOWERING PERIOD May–Oct.
HABITAT A wide range of grassland habitats, apart from on the most acidic soils; also in waste places and along roadsides.
FREQUENCY Very widespread and often abundant.

White Clover
■ *Trifolium repens* Up to 40cm

DESCRIPTION A creeping, hairless perennial plant that sends down roots where the leaf nodes touch the ground. The leaves are trefoil, the individual rounded leaflets having a whitish crescent pattern across the centre and translucent cross-veins. The delicately scented white flowers are arranged in loosely globular, long-stalked heads up to 2cm across. The flowers may occasionally be pink or purple.
FLOWERING PERIOD May–Oct.
HABITAT A wide range of grassland habitats, especially where there is grazing or trampling, and avoiding only the most waterlogged or acidic conditions.
FREQUENCY Widespread and very common (often planted in grass-seed mixtures), but scarce at high altitudes.

Wood Sorrel

■ *Oxalis acetosella* Up to 10cm

DESCRIPTION A creeping perennial plant with long-stalked trefoil leaves, green above and sometimes purple-tinged below, which often close up at night. The solitary white bell-shaped flowers are held up on long stalks and are about 1cm across. The delicate white petals usually have pink or purple veins.
FLOWERING PERIOD Apr–May.
HABITAT Damp, shady woodlands, conifer plantations, old hedgerows, mountain gullies and boulder slopes, and grykes in limestone pavements. Intolerant of dry or sunny conditions.
FREQUENCY Widespread and locally common in suitable habitats, but less common in E England and the Low Countries.

Meadow Crane's-bill

■ *Geranium pratense* Up to 75cm

DESCRIPTION A showy, clump-forming perennial with deeply divided, hairy leaves, cut into as many as 7 strongly toothed lobes. The 5-petalled blue flowers are up to 30mm across. When the flowers are fully open, the flower stalks are held erect and then droop until the long-beaked seeds have developed, when they stand upright once more. Crane's-bills are named after the fanciful resemblance of their long seed pods to a bird's bill.
FLOWERING PERIOD Jun–Sep.
HABITAT A variety of rough grassland habitats, mainly on calcareous soils, including verges, embankments and old meadows. Most common in lowland areas, but can be found on some sunny hill slopes.
FREQUENCY Locally common and widely naturalised in many lowland areas, but scarce or absent in the far NW and extreme SE.

Wood Crane's-bill
■ *Geranium sylvaticum* Up to 60cm

DESCRIPTION An eye-catching tufted, downy or hairy perennial with rounded leaves that are deeply divided into 5–7 lobes. The 5-petalled reddish-mauve flowers are up to 3cm across and held upright on long stalks. The fruits terminate in the long 'beak' typical of this family.
FLOWERING PERIOD Jun–Aug.
HABITAT Upland hay meadows, grassy banks, damp woodlands, mountain ledges and streamsides, thriving best in areas where there is little grazing by deer or cattle.
FREQUENCY Absent from S Britain and most of Ireland (where it is a great rarity), and also from the adjoining areas of France, Belgium and Holland. It is locally common in suitable habitats elsewhere.

Bloody Crane's-bill ■ *Geranium sanguineum* Up to 25cm

DESCRIPTION A clump-forming perennial with dark green leaves that are divided almost to the base into 5–7 lobes. The whole plant is downy, with slightly more hairy stems. The deep reddish-purple flowers are borne singly on short stalks, and are up to 30mm across. The

mature fruits terminate in the typical crane's-bill 'beak'.
FLOWERING PERIOD Jun–Aug.
HABITAT Lime-rich grasslands and open scrubby areas, sea cliffs, stabilised sand-dunes, limestone pavements and old quarries; also along some lanes and hedgerows as a garden escape.
FREQUENCY Locally common in some widely scattered, mainly coastal locations, especially on serpentine and dolerite rocks, but also inland as an established garden escape.

Herb Robert

■ *Geranium robertianum* Up to 30cm

DESCRIPTION A straggling, unpleasant-smelling annual with rather brittle stems and roughly triangular leaves, these deeply divided into 3–5 lobes, each of which is further divided pinnately. The 5-petalled pink flowers are up to 20mm across and have orange pollen showing at the centre. The beaked fruits are slightly wrinkled and hairy.
FLOWERING PERIOD Apr–Nov.
HABITAT Shade-tolerant and found in a wide range of habitats, including woodlands, hedgerows, shrubby gardens, waste ground, embankments, limestone pavements and coastal shingle. Not usually found on very acidic soils or at high altitudes.
FREQUENCY Widespread and very common.

Shining Crane's-bill

■ *Geranium lucidum* Up to 30cm

DESCRIPTION A short, branching, almost hairless annual with shiny rounded leaves, these cut almost to the centre; the leaves and stems sometimes become reddened later in the season. The 5-petalled pink flowers are up to 15mm across and have a slightly inflated calyx; the petals are not notched. The fruits are slender and hairless.
FLOWERING PERIOD May–Aug.
HABITAT Shady hedgerows, steep banks and scree slopes, old mortared walls, churchyards and derelict sites. Most common in limestone and chalk areas, and usually absent from very acidic conditions.
FREQUENCY Widespread but only locally common, and absent from high altitudes and the far N.

Common Stork's-bill
▪ *Erodium cicutarium* Up to 30cm

DESCRIPTION A stickily hairy annual plant, often with an unpleasant smell. The long leaves are finely divided, giving them a rather feathery appearance, and they have narrow stipules at the base. The 5-petalled pink flowers are up to 14mm across, but the petals are easily shed; these are followed by the typical long, beak-like fruits.
FLOWERING PERIOD May–Aug.
HABITAT Open, sunny, well-drained grasslands, very often near the sea. Dunes and heaths, embankments and wasteland are also favoured.
FREQUENCY Widespread and locally common, especially near the sea and in the SE. Absent from high altitudes and very wet areas.

Wood Spurge
▪ *Euphorbia amygdaloides* Up to 80cm

DESCRIPTION An upright, downy perennial plant with red stems and alternate long, narrow leaves. Snapped stems exude a milky latex that can stain human skin. The very small flowers have no petals or sepals, but are surrounded by bright yellow petal-like bracts, and are arranged in loose, rather flat umbels, making them very conspicuous in shady woodlands.
FLOWERING PERIOD Apr–May.
HABITAT Shady woodlands and deep lanes on neutral or slightly acidic soils, and occasionally on sheltered sea cliffs and gullies.
FREQUENCY Locally common in S England and Wales and N France, but very scarce or absent elsewhere.

Dog's Mercury
■ *Mercurialis perennis* Up to 35cm

DESCRIPTION A hairy, creeping, unbranched perennial with an unpleasant smell. The shiny oval leaves are arranged in pairs up the stem and have toothed margins. The tiny yellow flowers have no petals, and are arranged in open spikes with the separate sexes occurring on different plants. After pollination, the female flowers are succeeded by rounded, hairy fruits.
FLOWERING PERIOD Feb–Apr.
HABITAT Shady woodlands, hedgerows and gullies, limestone pavements and mountain ledges, generally in damp conditions but where the soil is free-draining.
FREQUENCY Widespread and very common in suitable habitats, even at quite high altitudes, but absent from the far N and very scarce in Ireland and Scandinavia.

Common Milkwort
■ *Polygala vulgaris* Up to 20cm

DESCRIPTION An erect or trailing, hairless perennial with narrow, unstalked alternate leaves. The small blue flowers are about 8mm long and are composed of 5 sepals, the inner 2 of which resemble large blue petals, and 3 smaller true petals. The stamens within the flower are also blue and slightly petal-like. Pink and white colour forms may also be found. The fruits are flat, heart-shaped and winged.
FLOWERING PERIOD May–Sep.
HABITAT Open grassland on hills, chalk downs, dunes and coastal cliffs. It tolerates a wide range of soil conditions, from heathland sands to moderately fertile loams.
FREQUENCY Widespread and common throughout.

Musk Mallow
■ *Malva moschata* Up to 75cm

DESCRIPTION A branched and rather hairy perennial with deeply dissected, alternate leaves. The showy rose-pink flowers, arranged in small clusters at the tips of leafy shoots, have 5 petals, which are very narrow at the base and have green sepals showing between them. The fruits are rounded, flat capsules.
FLOWERING PERIOD Jul–Aug.
HABITAT A variety of open grassy habitats, including roadsides, embankments, woodland clearings and waste ground. Well-drained soils seem to be preferred.
FREQUENCY Widespread and common, especially in lowland areas, but introduced and quite local in Scotland and Ireland, and locally common in S Norway and Sweden.

Common Mallow
■ *Malva sylvestris* Up to 1.5m

DESCRIPTION A rather variable perennial plant, which may push its way through other vegetation or grow in a compact, shrubby form. The leaves are rounded at the base of the plant, but 5-lobed on the stem; they may also show a small dark spot. The flowers are up to 45mm across and have 5 deep pink, purple-veined petals that are much longer than the sepals, which show between their narrow bases.
FLOWERING PERIOD Jun–Oct.
HABITAT Well-drained, sunny grassland areas, including roadsides, hedgerows, arable margins, cliff tops and neglected urban sites. An introduced and widely established species.
FREQUENCY Widespread and common over much of S Britain and N Europe, but absent from the far N and scarce in Scandinavia.

Slender St John's-wort

■ *Hypericum pulchrum* Up to 60cm

DESCRIPTION A hairless perennial plant with rounded stems and opposite, oval leaves that have translucent veins and perforations when viewed against the light; they may also have inrolled margins. The flowers open from orange-tipped buds, and the 5 yellow petals have tiny red dots and darker spots on the margins. The small seeds are produced inside smooth oval pods.
FLOWERING PERIOD Jul–Sep.
HABITAT Dry grassy places, heaths, woodland clearings and cliff tops where there is free-draining, slightly acidic soil.
FREQUENCY Very widespread and common, although more local in E areas.

Perforate St John's-wort

■ *Hypericum perforatum* Up to 80cm

DESCRIPTION An upright, hairless perennial plant with 2 narrow, raised ridges down the stem. The oval leaves are in opposite pairs, and show both translucent and black spots when held up to the light. The yellow flowers are up to 2cm across and the petals, and sometimes also the sepals, have black dots along the margins.
FLOWERING PERIOD Jul–Sep.
HABITAT A range of open, usually dry, grassy habitats, including hedgerows, embankments, woodland clearings and downland scrub.
FREQUENCY Widespread; commonest in the S, becoming scarce or absent in the far N.

Common Rock-rose
■ *Helianthemum nummularium* Up to 40cm

DESCRIPTION A much-branched, creeping, slightly woody perennial with opposite pairs of narrow, downy, untoothed leaves; the undersides of the leaves are white and the margins usually slightly inrolled. The flowers are up to 2.5cm across and the yellow petals look as if they have been crushed; in the centre of the flower is a dense ring of orange stamens.
FLOWERING PERIOD May–Sep.
HABITAT Open, sunny grasslands, usually well-grazed sites on chalky or limestone soils in the S. Also on more acidic but well-drained soils in the N.
FREQUENCY Locally common in the S of the region, but rather scarce and scattered elsewhere, and largely absent from Ireland and N Scandinavia.

Sweet Violet
■ *Viola odorata* Up to 15cm

DESCRIPTION A low-growing perennial with rooting runners and a tuft of long-stalked, rounded leaves, which become larger in summer. The leaves are downy, with toothed margins. The fragrant flowers are up to 15mm across and made up of 5 unequal petals, the lowest of which has a spur projecting back from the flower. The flowers are usually violet-blue, but can also be white or other delicate shades. Seeds are produced in egg-shaped capsules.
FLOWERING PERIOD Mar–May.
HABITAT Open woodlands, hedgerows, scrub, parks and churchyards, generally on lime-rich soils and in light shade.
FREQUENCY Widespread and locally common in the S and E of the region, becoming more local and scattered in the N.

Common Dog Violet
■ *Viola riviniana* Up to 12cm

DESCRIPTION An unscented violet with bluish-white flowers up to 25cm across that have a pale, stout spur with a notched tip, and pointed sepals. The heart-shaped leaves are almost hairless and have toothed stipules at the base.
FLOWERING PERIOD Mar–May.
HABITAT A wide variety of habitats, including woodlands, hedgerows, heathlands, cliff tops, mountain grasslands and gardens, where it can become a troublesome weed. Not usually found in very wet areas or on very acidic soils.
FREQUENCY Widespread and very common, and probably increasing in man-made habitats.

Wild Pansy
■ *Viola tricolor* Up to 12cm

DESCRIPTION A rather variable small annual or biennial, which can be either downy or hairless and can have violet, yellow or bicoloured flowers up to 25mm across. The leaves are long-stalked and lanceolate, and have large leaf-like stipules at the base. The pure yellow form is most likely to be encountered at coastal sites. An alternative name for this species is Heartsease, which is a reference to its former use as a medicinal herb.
FLOWERING PERIOD Apr–Nov.
HABITAT Sand-dune slacks, acid grassland on heaths and moors, cultivated ground and arable margins.
FREQUENCY Widespread, but only locally common and absent from large areas, especially in upland regions.

Field Pansy
■ *Viola arvensis* Up to 15cm

DESCRIPTION A short, rather variable, downy annual with long-stalked, oval, toothed basal leaves, and long leaf-like stipules up the flowering stem. The mostly white flowers are up to 15mm across and have an orange patch on the lowest petal; the sepals are the same length as the petals. Seeds are produced inside small, rounded capsules.
FLOWERING PERIOD Apr–Nov.
HABITAT Dry arable land, field margins and tracks, allotments and waste ground.
FREQUENCY Widespread and very common, and generally thought of as a weed of cultivation.

Purple Loosestrife
■ *Lythrum salicaria*
Up to 1.5m

DESCRIPTION A tall, downy perennial plant with showy spikes of purple flowers. The square-shaped stems have 4 raised lines on the surface, and the leaves are linear and arranged in opposite pairs. The flowers are deep reddish purple, up to 15mm across and have 6 petals.
FLOWERING PERIOD Jun–Aug.
HABITAT Damp grassland, especially pond and river margins and fens, and also some brackish habitats such as the upper reaches of estuaries.
FREQUENCY Widespread and locally common, but scarce in the N and in upland areas.

Enchanter's Nightshade

■ *Circaea lutetiana* Up to 65cm

DESCRIPTION A low-growing, patch-forming, downy perennial plant with opposite pairs of oval leaves with heart-shaped bases. Tall, leafless spikes of tiny white flowers arise above the leaves; the 2-petalled flowers are up to 8mm across, and once the petals have fallen the flower spike elongates. The tiny fruits have hooked bristles to aid seed dispersal.
FLOWERING PERIOD Jun–Sep.
HABITAT Woodland clearings, hedgerows, shady gardens and churchyards.
FREQUENCY Widespread and common in the S, where it is becoming increasingly common as a garden weed, and scarce or absent in the N.

Rosebay Willowherb

■ *Chamaenerion angustifolium* Up to 1.5m

DESCRIPTION A tall, showy and very familiar perennial plant, often forming dense patches on newly cleared sites. The long, lanceolate leaves are arranged spirally up the stem, and this is topped by the dense spike of pinkish-purple 4-petalled flowers, which open from the bottom of the spike upwards. The flowers are followed by long seed pods, which split to reveal the cottony seeds.
FLOWERING PERIOD Jul–Sep.
HABITAT Roadsides, hedgerows, embankments, waste ground and urban sites, plus heathlands and mountain slopes.
FREQUENCY Widespread and very common in a range of habitats, but only locally common in the far NW.

Great Willowherb

■ *Epilobium hirsutum* Up to 2m

DESCRIPTION A tall, downy perennial with a round stem and hairy, oval, opposite leaves, the upper ones clasping the stem. The pinkish-purple flowers, with a paler centre, are up to 2.5cm across and have a white 4-lobed stigma in the centre. They grow in loose, open spikes and are followed by long seed pods, which split to release the cottony, wind-dispersed seeds.
FLOWERING PERIOD Jul–Aug.
HABITAT Damp grassy areas, including river and pond margins, fens, sheltered lanes and woodland glades.
FREQUENCY Widespread and common in the S, but absent from the far N.

Common Evening Primrose

■ *Oenothera biennis* Up to 1.5m

DESCRIPTION An upright, downy biennial plant with large, showy, 4-petalled yellow flowers, up to 6cm across, that become fragrant at dusk and are very short-lived once open. The large, lanceolate leaves have prominent red veins. Several other very similar species, occur and hybrids are common and very confusing.
FLOWERING PERIOD Jun–Sep.
HABITAT Sand-dunes, dry grasslands, roadsides, embankments and waste places.
FREQUENCY Introduced from North America and now scattered across a wide area of the S and E of the region; apparently increasing in range and frequency.

Dwarf Cornel

■ *Cornus suecica* Up to 15cm

DESCRIPTION A short, creeping perennial plant with woody lower stems and opposite pairs of unstalked oval leaves that have prominent veins. The tiny, dark purple flowers are only about 3mm across and are arranged in a tight umbel, but they are offset by 4 large creamy-white petal-like bracts up to 15mm across. Not all plants in a patch will produce flowers. The fruits are clusters of small red berries.
FLOWERING PERIOD Jun–Aug.
HABITAT Upland grassland, moors and mountain slopes, sometimes under Bracken or heather, but not on limestone rocks or soils.
FREQUENCY Locally common in uplands of Scotland and a few scattered localities in N England. Absent from Wales and Ireland, but widespread and common in Scandinavia.

Sanicle

■ *Sanicula europaea* Up to 50cm

DESCRIPTION A slender, low-growing, hairless perennial plant with long-stalked 3–5-lobed leaves that have toothed margins. The very small, 5-petalled pinkish-white flowers are grouped together in tiny umbels on a single long stalk held well above the leaves. The tiny egg-shaped fruits have short, hooked bristles to aid dispersal.
FLOWERING PERIOD May–Jul.
HABITAT Shady areas in deciduous woodlands, usually on lime-rich or heavy soils. In some S Beech woods it may form a complete carpet.
FREQUENCY Widespread and locally common, becoming scarce in the N and at high altitudes.

Sea Holly

■ *Eryngium maritimum* Up to 60cm

DESCRIPTION A very distinctive, showy, hairless perennial plant with extremely tough, spiny blue-green leaves bearing a fanciful resemblance to Holly. The leaves clasp the tough, solid stem and have prominent white margins and white veins. The flowers are compacted into a globular blue umbel up to 4cm long with a basal ruff of more spiny bracts.
FLOWERING PERIOD Jun–Sep.
HABITAT Just above the high-tide line on coastal shingle and sandy shores where there is not too much trampling or wave action.
FREQUENCY Very local and restricted to coasts in the S and SW.

Cow Parsley

■ *Anthriscus sylvestris* Up to 1m

DESCRIPTION A tall, slightly downy herbaceous perennial with hollow, sometimes purple stems. The large leaves are pinnately divided and rather fern-like in form. They usually appear well before the flowers. The tiny white 5-petalled flowers are arranged in umbels up to 6cm across, many of which are supported by each plant.
FLOWERING PERIOD Apr–Jun.
HABITAT Hedgerows, embankments, woodland margins and clearings; tolerant of light shade.
FREQUENCY Often very common along lanes and roadsides, producing a continuous white haze of blossom. Widespread, becoming less common only in the far N.

Hogweed

■ *Heracleum sphondylium* Up to 2.5m

DESCRIPTION A robust perennial or biennial with hollow, ridged stems and large, pinnately divided leaves that are up to 60cm long with very broad, toothed leaflets. The white flowers are arranged in umbels about 20cm across, and the outer flowers have 1 petal that is much larger than the others. The flowers smell strongly of pigs, hence the plant's common name. Large, flattened, hairless seeds are produced in great abundance.
FLOWERING PERIOD Apr–Nov.
HABITAT Meadows, open woodlands, roadsides, hedgerows, derelict agricultural land and urban sites.
FREQUENCY Widespread and, in places, very common; a familiar sight on roadsides everywhere.

Hemlock

■ *Conium maculatum* Up to 2m

DESCRIPTION A tall, hairless and extremely poisonous perennial plant with hollow stems covered in purple blotches. The leaves are very finely divided into numerous narrow leaflets, and when crushed they exude an unpleasant smell. The small white flowers form umbels about 5cm across, many of which are supported by each plant.
FLOWERING PERIOD Jun–Aug.
HABITAT Damp, open grassland, roadsides, embankments, river and pond margins. Often rapidly colonises newly created roadside verges.
FREQUENCY Common in the S and E, but scattered and more local elsewhere and absent from the far N.

Wild Angelica

■ *Angelica sylvestris* Up to 2m

DESCRIPTION A tall, robust, almost hairless perennial with distinctive hollow, ridged stems tinged with purple. The leaf stalks have inflated sheath-like bases that clasp the stem, and the leaves are divided into broad, toothed leaflets. The tiny white, sometimes pink-tinged, 5-petalled flowers are grouped into large, domed umbels up to 15cm across, and the seeds are oval, flattened and winged. FLOWERING PERIOD Jul–Sep. HABITAT Damp grassland, woodland rides, pond and river margins, fens and shaded gullies. FREQUENCY Very widespread and locally common in suitable habitats.

Ground Elder

■ *Aegopodium podagraria* Up to 60cm

DESCRIPTION A hairless perennial with a creeping, patch-forming habit, enabling it to become a persistent and resilient garden weed. The light green triangular leaves are 3-lobed and each lobe is further divided into 3 smaller lobes. The white flowers form compact, rounded umbels up to 6cm across. The fruits are egg-shaped and have a ridged surface. This plant can spread by both seed dispersal and runners. FLOWERING PERIOD Jun–Aug. HABITAT Damp, shady woodlands, hedgerows, roadsides and gardens. FREQUENCY Very widespread and probably increasing in abundance as it spreads from garden to garden. Probably not native to Britain, but introduced and dispersed by horticulture. Scarce or absent from N Scandinavia.

Wild Parsnip

■ *Pastinaca sativa* Up to 1m

DESCRIPTION A tall, roughly hairy and strong-smelling perennial plant with pinnately divided leaves made up of toothed, oval leaflets. The flowers are bright yellow, an unusual colour in umbellifers, and grouped in open umbels up to 9cm across. The seeds are a flattened egg shape and have winged margins. The **Garden Parsnip** P. *sativa* var. *hortensis* is closely related and very similar, and sometimes becomes established in the wild.
FLOWERING PERIOD Jun–Sep.
HABITAT Open grassland areas, normally on well-drained lime-rich soils such as chalk downlands and roadside embankments.
FREQUENCY Locally common in the S and E on suitable soils, but scarce or absent elsewhere.

Hemlock Water-dropwort

■ *Oenanthe crocata* Up to 1.25m

DESCRIPTION A tall, spreading, hairless and very poisonous perennial plant with a strong smell of parsley when bruised. The hollow stems are grooved, and the leaves are much divided into small, toothed, tapering segments. The white flowers form compact, domed umbels up to 10cm across, and the fruits are cylindrical.
FLOWERING PERIOD Jun–Aug.
HABITAT Damp grassy places, ditches, river and pond margins. Mainly in lowland areas.
FREQUENCY Locally common in the W, sometimes dominating wet habitats, but scarce or absent elsewhere.

Rock Samphire ■ *Crithmum maritimum* Up to 40cm

DESCRIPTION A low-growing, much-branched perennial plant, greyish green in colour, with leaves divided into numerous long, narrow, fleshy lobes, triangular in cross section and strong-smelling when crushed. The creamy-yellow flowers form umbels up to 6cm across;

they are composed of as many 30 rays and have small leafy bracts below.
FLOWERING PERIOD Jun–Sep.
HABITAT Coastal rocks and shingle, and sometimes man-made habitats like jetties and breakwaters, often very close to the high-tide line and likely to be splashed by the waves.
FREQUENCY Locally common in suitable habitats around S and W coasts.

Wild Carrot ■ *Daucus carota* ssp. *carota* Up to 75cm

DESCRIPTION A roughly hairy, upright or rather bushy perennial with solid, ridged stems. The leaves are pinnately divided into many narrow leaflets. The tiny white flowers are grouped into dense, convex umbels, up to 7cm across, with a single small red flower in the centre and divided bracts below. When the oval, spiny seeds form, the umbel then

becomes concave. The very similar **Sea Carrot** *Daucus carota* ssp. *gummifer* has flat or convex umbels in fruit.
FLOWERING PERIOD Jun–Sep.
HABITAT Rough, open grassland, generally on dry chalky soils, especially downlands, steep embankments and roadside verges. Sea Carrots are common on sunny sea cliffs and coastal grasslands.
FREQUENCY Locally common in suitable habitats in the S and E, becoming scarce or absent in the N.

Primrose

■ *Primula vulgaris* Up to 20cm

DESCRIPTION A familiar and attractive low-growing, hairy perennial plant with long, wrinkled leaves tapering to the base and forming a clump-like basal rosette. The delicately scented, solitary yellow flowers, up to 30mm across, are borne on long, hairy pink flower stalks and usually have a darker centre. Flowers are either 'pin-eyed', in which the stigma is prominent in the centre, or 'thrum-eyed', in which the anthers are prominent. Seeds are produced inside smooth capsules.
FLOWERING PERIOD Feb–May.
HABITAT Woodlands, hedgerows, sea cliffs, mountain slopes and gullies; also in gardens, where they have often been collected from wild stock. Spreading rapidly onto new roadside embankments.
FREQUENCY Very widespread and sometimes locally abundant.

Cowslip

■ *Primula veris* Up to 25cm

DESCRIPTION An attractive downy perennial plant with long, tapering, wrinkled leaves forming a basal rosette. The pleasantly scented yellow flowers are up to 15mm across and bell-shaped; they are usually grouped in 1-sided umbels of up to 30 flowers, which are a richer colour than Primroses and have an orange centre. Cowslips may hybridise with Primroses to give rise to the **False Oxlip** *P. veris* × *vulgaris*.
FLOWERING PERIOD Apr–May.
HABITAT Open, sunny and often well-grazed grassland, especially chalk downland and embankments that are S-facing. Rapidly colonising roadside verges.
FREQUENCY Widespread in the S and E, and locally common there in suitable habitats, but scarce or absent in large parts of the N and NW.

Oxlip

▪ *Primula elatior* Up to 20cm

DESCRIPTION A low-growing, hairy perennial plant with long, wrinkled leaves that taper abruptly towards the base; they are arranged in a loose rosette around the tall flower stalk. The pale yellow unscented flowers are arranged in a drooping, 1-sided cluster; each flower is about 20mm across.
FLOWERING PERIOD Apr–May.
HABITAT Damp, lightly shaded woodlands and scrub on heavy clay soils, usually in the absence of Primroses.
FREQUENCY Locally common only in the E of England in suitable woodland habitats, where it may form extensive patches, but much more common in N Europe.

Yellow Loosestrife

▪ *Lysimachia vulgaris* Up to 1m

DESCRIPTION An attractive upright, downy perennial with whorls of broadly lanceolate leaves up the stem; the leaves are often marked with black or orange dots. The bright yellow 5-petalled flowers, marked with pale spots, are up to 15mm across and grouped together in loose, branched, rather leafy clusters. Seeds are produced in small capsules.
FLOWERING PERIOD Jul–Aug.
HABITAT Riverside meadows, fens, lake margins and wet grassland.
FREQUENCY Widespread in the S and E, and locally common in suitable habitats, but declining in some areas owing to habitat loss.

Yellow Pimpernel

■ *Lysimachia nemorum* Creeping

DESCRIPTION A creeping, hairless perennial with pairs of oval or heart-shaped, short-stalked leaves; despite their delicate appearance, they seem to resist grazing by slugs and snails. The 5-petalled, star-shaped, bright yellow flowers are up to 12mm across and grow on slender stalks arising from the leaf axils; they open fully only in fine, sunny weather. Seeds are produced in small greenish capsules. FLOWERING PERIOD May–Aug. HABITAT Damp, shady woodlands, hedgerows and rocky areas. FREQUENCY Very widespread and can be quite common in suitable places.

Scarlet Pimpernel

■ *Lysimachia arvensis* Creeping

DESCRIPTION A creeping, mostly hairless perennial, with square stems and opposite pairs of unstalked oval, pointed leaves that have black dots on the underside. The 5-petalled flowers are up to 15mm across and are usually orange with a purple centre, but may sometimes be pink, or occasionally even lilac or blue. The flowers grow on slender stalks that arise from the base of the leaves, and open fully only in bright sunshine, hence the plant's old rural name of Poor Man's Weather Glass. FLOWERING PERIOD May–Oct. HABITAT Gardens, waste ground, dry tracks, sand-dunes and urban sites. FREQUENCY Common in the S and E, becoming scarce or absent elsewhere.

Bog Pimpernel
■ *Lysimachia tenella* Creeping

DESCRIPTION A most attractive and delicate mat-forming perennial plant, with a mass of creeping reddish stems bearing pairs of oval, untoothed, opposite leaves. The bell-shaped pink flowers are up to 10mm long and are raised up singly on slender stalks arising from the leaf bases. The flowers open best in full sun, and in very shaded habitats may not be produced at all.
FLOWERING PERIOD May–Sep.
HABITAT Bogs, wet meadows, dune slacks and fens, usually on peaty soils and in unshaded situations.
FREQUENCY Locally common in the W, but scattered and very scarce elsewhere.

Chickweed Wintergreen ■ *Lysimachia europaea* Up to 20cm

DESCRIPTION A rather delicate, hairless perennial plant with a single whorl of oval leaves surrounding the stem well above ground level. The pure white star-like flowers are up to

15mm across and usually have about 7 petals; they arise from the leaf rosette on slender stalks and are mostly solitary.
FLOWERING PERIOD Jun–Jul.
HABITAT Mossy ground in conifer woods, and shady places on heather moors and heaths.
FREQUENCY Locally common only in suitable habitats in Scotland and N England, abundant in Scandinavia, but absent from the S and W.

Ling ▪ *Calluna vulgaris* Up to 60cm

DESCRIPTION A dense, low-growing evergreen shrub, also known as Heather. The stems in mature plants are gnarled and woody, but young shoots are greener and covered with closely packed, small, narrow leaves arranged in opposite rows. The leaves usually have inrolled margins and are often covered with a grey down. The tiny, 4-petalled pinkish-purple flowers are only about 3–4mm across, growing in tightly packed, erect spikes.
FLOWERING PERIOD Jul–Sep.
HABITAT Upland moors, lowland heaths, acidic grasslands and other open habitats on acid soils, but not in completely waterlogged sites.
FREQUENCY Widespread and very common, but only locally common in some E areas. Huge tracts of moorland become purple with Ling flowers in late summer.

Bell Heather
▪ *Erica cinerea* Up to 60cm

DESCRIPTION A low-growing and tough undershrub with wiry stems and narrow, glossy green leaves arranged in whorls of 3 along the stem. The reddish-purple bell-shaped flowers are about 5mm long and grouped in loose spikes or heads. Bell Heather often grows with other heather species but its bright colours make it stand out.
FLOWERING PERIOD May–Sep.
HABITAT Drier parts of heaths and moors, sometimes overlapping with Ling (above), and dry, sunny mountain slopes, cliff tops and acid grassland.
FREQUENCY Widespread and common in suitable habitats over a large area of Britain, Ireland and W France, but scarce or absent further E.

Cross-leaved Heath

■ *Erica tetralix* Up to 50cm

DESCRIPTION A short, rather greyish undershrub with brittle woody stems and greyish-green needle-like leaves arranged in whorls of 4 around the stems. The globular rose-pink flowers are up to 7mm long and grouped into small, compact, slightly 1-sided heads at the top of longer leafy shoots.
FLOWERING PERIOD Jun–Oct.
HABITAT Grows best in the wettest parts of heaths and moors, often alongside streams and sphagnum bogs, replacing the other heather species in wet conditions.
FREQUENCY Widespread and locally common in suitable habitats. Very scarce or absent in the E of the region and N Scandinavia.

Bilberry

■ *Vaccinium myrtillus* Up to 75cm

DESCRIPTION A much-branched, hairless deciduous undershrub with angled green twigs. The bright green leaves are oval and slightly toothed. The bell-shaped flowers are pale pink, sometimes with a greenish tinge, and usually solitary along the leafy shoots. The fruits are an edible and succulent black berry with a purple bloom that is easily brushed off when they are handled. Known as Blaeberry in Scotland and Whortleberry in S England.
FLOWERING PERIOD Apr–Jun.
HABITAT Confined to acid soils in conifer woods, moorlands and heathlands, sometimes reaching quite high altitudes and growing in exposed rocky situations.
FREQUENCY Widespread and very common in suitable habitats. Absent only from very wet sites or lime-rich soils, and mostly scarce or absent from central and E England.

Cowberry

■ *Vaccinium vitis-idaea* Up to 30cm

DESCRIPTION A straggling evergreen undershrub with rounded twigs that are downy near the growing tips. The oval, glossy green leaves have untoothed margins, which are rolled downwards, and a slightly notched tip. The pale pink flowers are about 5–8mm long and bell-shaped, and have a more open mouth than those of Bilberry (p. 78). They are usually grouped in small, nodding clusters. The fruits are shiny red berries. Occasional hybrids with Bilberry may be found.
FLOWERING PERIOD May–Jun.
HABITAT Moors, heaths, mountain slopes and open woodlands on acid soils.
FREQUENCY Locally common from N Wales northwards, but quite local and scarce in Ireland and very scarce in the S of the region.

Cranberry

■ *Vaccinium oxycoccos* Up to 30cm

DESCRIPTION A low-growing, straggling evergreen undershrub with numerous wiry stems. Narrow, pointed, oval leaves are arranged alternately along the stems; their margins are inrolled and they are usually very pale on the underside. The flowers, which grow on very slender stalks, have 4 bright pink petals that are turned back to reveal a cluster of 8 yellow stamens. The fruit is the highly prized round to oval red berry of culinary importance.
FLOWERING PERIOD Jun–Jul.
HABITAT Bogs and very wet heaths, often growing on cushions of sphagnum moss.
FREQUENCY Locally common only in suitable habitats and widely scattered locations; most likely to be found from N Wales northwards. Very scarce in the S of the region.

Crowberry
▪ *Empetrum nigrum* Creeping

DESCRIPTION A mat-forming evergreen undershrub with trailing woody stems that may reach out for up to 1m. The narrow, pointed and rather shiny leaves are arranged spirally along the stem; their margins are inrolled downwards. The tiny, 6-petalled, pale pink flowers grow out the leaf axils, and male and female flowers are found on separate plants. The fruits are small berries that turn from green, through pink and purple, to shiny black.
FLOWERING PERIOD Apr–Jun.
HABITAT Moors, boggy areas and mountain slopes, often reaching very high altitudes.
FREQUENCY Locally common in suitable upland habitats in N Britain, and closer to sea-level in the extreme N.

Round-leaved Wintergreen
▪ *Pyrola rotundifolia* Up to 15cm

DESCRIPTION A low-growing, hairless perennial plant with long-stalked, rounded leaves forming a loose basal rosette. The flowers grow on a single long spike; each 5-petalled white flower is about 12mm across and has a single S-shaped style in its centre that protrudes quite prominently from between the petals. Seeds are produced in small capsules.
FLOWERING PERIOD Jun–Sep.
HABITAT Damp, partly shaded sites on calcareous soils in dunes, open woodland, rock ledges and fens.
FREQUENCY On a few scattered sites in the S, becoming more common in the N and E of the region.

Thrift

■ *Armeria maritima* Up to 30cm

DESCRIPTION A low-growing, clump-forming perennial plant with prolific showy flower heads. The narrow, rather fleshy, dark green leaves form compact springy cushions attached by a deep tap root. The globular flower heads are supported on downy stalks that arise from the leaf cushion. Beneath the pink flowers are brown papery bracts. Darker pink, or sometimes pure white, colour forms occasionally appear.
FLOWERING PERIOD Apr–Jul.
HABITAT Sea cliffs, stabilised shingle and saltmarshes, and some mountain ledges well inland.
FREQUENCY Widespread and locally common around the entire coast, with a few inland sites.

Common Sea Lavender

■ *Limonium vulgare* Up to 30cm

DESCRIPTION A short, hairless perennial plant, completely unrelated to the true Lavender, with lanceolate, long-stalked deciduous leaves. The small lavender-coloured flowers, which are unscented, grow in a long, curved, leafless spike. Each flower is about 6mm long, and the flower heads are about 8cm long. Growing in dense masses, they create a most attractive sight in saltmarshes.
FLOWERING PERIOD Jul–Sep.
HABITAT Upper reaches of saltmarshes, in sheltered areas regularly inundated by the tides.
FREQUENCY Widespread around S and E coasts, more local and scattered elsewhere, and absent from the far N.

Common Centaury

■ *Centaurium erythraea* Up to 25cm

DESCRIPTION A hairless annual or
biennial plant with a basal rosette
of narrow, pointed oval leaves, and
further narrow leaves up the stem.
The 5-petalled, star-shaped pink
flowers, up to 15mm across, grow
in small clusters on numerous side
shoots, and have yellow anthers. Very
occasionally, a white form appears.
FLOWERING PERIOD Jun–Oct.
HABITAT Dry grassy areas, usually
in full sun, on chalk downs, roadside
verges, embankments, cliff tops and
sand-dunes.
FREQUENCY Locally common in
the S, becoming scarce in the N, and
absent from N areas of Scandinavia.

Yellow-wort

■ *Blackstonia perfoliata* Up to 35cm

DESCRIPTION An upright, hairless
grey-green annual plant with
opposite pairs of rather waxy-looking
pointed oval leaves that clasp the
stem, and a basal rosette of similar
oval leaves, which are separated at
the base. The eye-catching bright
yellow flowers, which open only in
good sunlight, are up to 15mm across
and are made up of 6–8 petals.
FLOWERING PERIOD Jun–Oct.
HABITAT Dry, open grassland
on lime-rich soils, especially on
chalk downs and sand-dunes, and
sometimes on roadside verges and
embankments.
FREQUENCY Locally common in
suitable habitats in the S of the
region, very scarce in the far W and
absent from the N.

Marsh Gentian

■ *Gentiana pneumonanthe* Up to 30cm

DESCRIPTION A very attractive hairless
perennial with bright blue trumpet-shaped flowers.
The flowers are up to 4cm long and have 5 vertical
green stripes on the outside; the individual flowers
are clustered in terminal heads. The narrow leaves
have a single central vein and are found in opposite
pairs up the stem. This is a very conspicuous
species when the flowers are fully open in sunshine,
but in shady conditions, and outside the flowering
period, it is very difficult to locate and is easily
overlooked.
FLOWERING PERIOD Jul–Oct.
HABITAT Damp grassy heaths and boggy areas
on acidic soils.
FREQUENCY Only locally common in scattered
sites in the S and E of the region, and declining
in many areas through habitat loss and drainage.

Autumn Gentian

■ *Gentianella amarella* Up to 25cm

DESCRIPTION A low-growing,
hairless biennial plant, often with
a purple tinge to the foliage. The
narrow, pointed leaves form a basal
rosette in the 1st year, but often
wither away when the leafy flower
spike appears in the 2nd year.
The purple flowers may have 4 or
5 pointed lobes and are densely
clustered in upright shoots. The
calyx teeth are sharply pointed and
more or less equal in length.
FLOWERING PERIOD Jul–Oct.
HABITAT Chalk and limestone
grassland, dunes and cliff tops,
embankments and dry spoil tips.
FREQUENCY Widespread and locally
common, especially in the S and in
lowland areas.

Bogbean

■ *Menyanthes trifoliata* Up to 20cm

DESCRIPTION A creeping, hairless aquatic plant with distinctive 3-lobed leaves that bear a resemblance to **Broad Bean** *Vicia faba* leaves; they are borne on long, partly submerged runners that may spread for over 1m. The star-shaped pinkish-white flowers are up to 15mm across and have 5 fringed petal lobes.
FLOWERING PERIOD Mar–Jun.
HABITAT Shallow water, such as pond and canal margins, damp peaty ground and fens; always in sunny locations.
FREQUENCY Widespread and locally very common in suitable habitats.

Lady's Bedstraw

■ *Galium verum* Up to 30cm

DESCRIPTION A sprawling, mostly hairless perennial plant with whorls of narrow, pointed leaves that smell of new-mown hay when crushed. The tiny, 4-petalled yellow flowers, only 3mm across, grow in dense clusters, and the whole plant often scrambles through other low vegetation, making it very conspicuous in summer. Bedstraws were once collected and dried for use as human bedding, hence their common name.
FLOWERING PERIOD Jun–Sep.
HABITAT Dry, open grassland habitats, especially on lime-rich soils.
FREQUENCY Widespread and often very common.

Hedge Bedstraw
■ *Galium album* Up to 1.5m

DESCRIPTION A scrambling perennial plant
with whorls of 6–8 narrow, pointed leaves
at intervals around the smooth, square stem.
The very small white flowers are only about
4mm across and have 4 sharply pointed petals;
they are arranged in large, open clusters. The
rounded fruits have a wrinkled texture.
FLOWERING PERIOD Jun–Sep.
HABITAT Hedgerows and dry grassy places,
preferring lime-rich soils and plenty of
sunshine.
FREQUENCY Widespread and common in
the S and E of the region, becoming more
scattered in the N and NW, and introduced
in a few areas in Ireland.

Common Cleavers
■ *Galium aparine* Up to 1.5m

DESCRIPTION A rather prickly, sprawling
annual plant that clambers over other
vegetation and clings to passing animals
and humans. The square stems are covered
with many backward-pointing bristles and
the narrow, pointed leaves also have bristles
on their tips. The flowers are very small and
inconspicuous, rarely more than 2mm across,
and are a pale greenish white. The rounded
fruits have numerous hooked bristles to aid
seed dispersal by animals.
FLOWERING PERIOD May–Sep.
HABITAT Cultivated land, hedgerows,
grasslands, waste
places, scree slopes
and embankments.
FREQUENCY
Widespread and very
common, except in
upland areas.

seeds and flowers

Crosswort
■ *Cruciata laevipes* Up to 50cm

DESCRIPTION A low-growing, tufted perennial plant with softly hairy, square stems and whorls of 4 oval, 3-veined leaves, from which the tiny yellow flowers arise in dense clusters. Each flower is only about 3mm across, consisting of 4 pointed yellow petals, but they are grouped in dense clusters. A delicate scent can be detected at close quarters.
FLOWERING PERIOD Apr–Jun.
HABITAT Grassy places, hedgerows, scrub and open, sunny woodland rides, usually on neutral or lime-rich soils.
FREQUENCY Widespread and common in the S and NE, but more scattered elsewhere, and absent from the far N and high altitudes.

Common Dodder
■ *Cuscuta epithymum* Climbing

DESCRIPTION A curious leafless parasitic plant that consists almost entirely of slender, twining red stems. The stems envelop the host plant, from which it obtains its nutrition, and invariably climb in an anticlockwise direction. The small pink flowers are only 3–4mm across, but grow in dense clusters at intervals along the stem, and produce a light scent.
FLOWERING PERIOD Jul–Sep.
HABITAT Heaths, cliff tops, chalk downs, open scrub areas and other sites where the usual host plants (mainly Gorse, p. 47, and heathers) occur.
FREQUENCY Scattered and locally common in the S, but scarce or absent further N; probably declining.

Hedge Bindweed

■ *Calystegia sepium*
Climbing, up to 2–3m

DESCRIPTION A fast-growing, hairless perennial that sends twining stems over other plants in order to gain the best growing position. The large, trumpet-shaped white flowers can be up to 6cm across; at their base are 2 large leaf-like bracts that half cover the sepals. The leaves are arrow-shaped and hairless.
FLOWERING PERIOD Jun–Sep.
HABITAT Cultivated areas, hedgerows, woodland margins, roadsides, urban areas and waste ground.
FREQUENCY Widespread and sometimes common enough to be a serious garden weed. Absent from the extreme N.

Field Bindweed

■ *Convolvulus arvensis*
Creeping or climbing, up to 3m

DESCRIPTION An attractive but troublesome perennial plant that wraps itself around other plants in order to gain height. The scented, funnel-shaped flowers are up to 3cm across and are usually pink with white stripes, but they can sometimes be pure white. The leaves are arrow-shaped and have long stalks. A deep root system enables the plant to survive gardening operations.
FLOWERING PERIOD Jun–Sep.
HABITAT Gardens, cultivated areas, waste ground, hedgerows, embankments and urban areas.
FREQUENCY Widespread and common in the S, very scarce or absent in the far N.

Sea Bindweed

■ *Calystegia soldanella* Creeping

DESCRIPTION A hairless, creeping perennial plant with long stems spreading for up to 1m, and bearing long-stalked, fleshy, kidney-shaped leaves. The attractive funnel-shaped flowers are up to 5cm across, are pink with 5 white stripes, and are raised from the stem on slender stalks. FLOWERING PERIOD Jun–Aug. HABITAT Among Marram Grass *Ammophila arenaria* on sand-dunes, and sometimes on stabilised shingle beaches among Sea Holly (p. 68), where there is no trampling. FREQUENCY Widespread and locally common around most of the coastline, but declining on heavily used shorelines.

Common Comfrey

■ *Symphytum officinale* Up to 1m

DESCRIPTION A roughly hairy perennial plant, usually found growing in large colonies. The stems are winged, and the leaves are oval and very hairy; their stalks run down the main stem towards the next leaves. The bell-shaped flowers, arranged in drooping clusters, are up to 18mm long and have a protruding style. They may be creamy white or pink to purple. FLOWERING PERIOD May–Jun. HABITAT Damp, slightly shaded areas, beside rivers or ditches, often on woodland rides on heavy soils. FREQUENCY Widespread and common in the S and E, but becoming more scattered and local in the N and W, and absent from high altitudes.

Hound's-tongue

■ *Cynoglossum officinale* Up to 75cm

DESCRIPTION An upright, softly downy perennial plant with a greyish appearance; it smells distinctly of mice when the long, narrow leaves are squeezed. The maroon flowers have 5 petal lobes, are about 6mm across and grow in small clusters. The fruits are groups of 4 flattened nutlets that are covered in hooked bristles to aid dispersal by animals.
FLOWERING PERIOD Jun–Aug.
HABITAT Dry, open, disturbed ground, especially Rabbit warrens on chalk soils, sand-dunes and sunny field margins.
FREQUENCY Locally common in scattered locations in the S and E, but scarce or absent elsewhere.

Common Gromwell

■ *Lithospermum officinale* Up to 60cm

DESCRIPTION An upright, rather hairy perennial plant, often with numerous leafy side branches and narrow, unstalked leaves showing prominent veins. The small creamy-white flowers, usually less than 4mm across, have 5 petal lobes and grow in small terminal clusters. The fruits are small, shiny white nutlets, appearing to be made of porcelain, and persisting long after the foliage has withered.
FLOWERING PERIOD Jun–Jul.
HABITAT Woodland and field margins, hedgerows and scrub, mostly in sunny situations and on well-drained calcareous soils.
FREQUENCY Locally common in the S, and scattered in lowland areas in the E, but scarce in the W and largely absent in the N of the region.

Viper's Bugloss

■ *Echium vulgare* Up to 80cm

DESCRIPTION An upright biennial plant covered in strong reddish-hued bristles. The basal leaves are stalked, while the stem leaves are unstalked, narrow and pointed. The stunning blue funnel-shaped flowers have protruding purple stamens and are up to 2cm long. They are arranged in tall, showy spikes, often with several plants together forming a conspicuous patch.
FLOWERING PERIOD May–Sep.
HABITAT Open, sunny sites on dry soils, such as chalk downs, dunes, roadside verges, embankments, quarries, and waste or disturbed ground.
FREQUENCY Widespread and locally common in the S and E, but more scattered elsewhere and not found at high altitudes.

Water Forget-me-not

■ *Myosotis scorpioides* Up to 20cm

DESCRIPTION A creeping, downy perennial plant with upright, slightly angular flowering stems bearing close-pressed hairs. The unstalked leaves are narrow and oblong. The 5-lobed sky-blue flowers are up to 1cm across and have a yellow centre that acts as a honey guide for bees; pink flowers are occasionally found. The flowers grow in arched clusters held above the leaves.
FLOWERING PERIOD Jun–Sep.
HABITAT A variety of watery habitats, including pond and stream margins, fens and marshes, usually on fertile soils. May occasionally be found forming floating mats of vegetation.
FREQUENCY Widespread and locally common in most areas.

Tufted Forget-me-not
■ *Myosotis laxa* Up to 12cm

DESCRIPTION A much-branched, hairy annual or biennial plant, which lacks runners but is otherwise rather similar to the Water Forget-me-not (p. 90). The round stems and oblong leaves have close-pressed hairs. The blue flowers are up to 4mm across, and have rounded lobes and pointed calyx teeth; they form slightly elongated clusters. The fruits are produced on long stalks.
FLOWERING PERIOD May–Aug.
HABITAT Damp grassland, pastures, woodland rides, stream and pond margins, and fens; often grows alongside Water Forget-me-not, with which it can form hybrids.
FREQUENCY Widespread and very common, but not found at high altitudes.

Bugle ■ *Ajuga reptans* Up to 20cm

DESCRIPTION A small, upright perennial plant with square stems, which are hairy on 2 opposite sides only. The leaves are oval, sometimes with a bronze tinge; the lower ones are stalked and the upper ones unstalked. Leafy runners grow out from the base of the plant. Leaf-like bracts grow between the whorls of flowers in a tapering flower spike. The bluish-violet bugle-shaped flowers are about 15mm long and have a divided lower lip that usually has paler veins.
FLOWERING PERIOD Apr–May.
HABITAT Damp woodland rides, shady lanes and banks, cliffs, screes and old pastures, preferring neutral or acidic soils.
FREQUENCY Widespread and very common in suitable habitats, and absent only from the extreme N.

Skullcap

■ *Scutellaria galericulata* Up to 40cm

DESCRIPTION A low-growing or sometimes sprawling perennial with slightly aromatic leaves that may be either smooth or downy. The oval, toothed leaves grow on short stalks. The bright blue flowers grow in opposite pairs on upright, leafy flower spikes; each flower can be up to 15mm long. The common name derives from the cap-like appearance of the upper part of the flower.
FLOWERING PERIOD Jun–Sep.
HABITAT A range of damp grassy habitats, including river and pond margins, fens and marshes, rushy meadows, dune slacks and wet woodlands.
FREQUENCY Widespread and locally quite common, except in the far N.

Self-heal

■ *Prunella vulgaris* Up to 20cm

DESCRIPTION A low-growing, downy perennial with creeping stems that send down rootlets at intervals. The unstalked, oval leaves grow in opposite pairs on the squarish stems, and the flowers are borne on upright flowering stems. Individual flowers are up to 15mm long and a deep blue-violet colour; they are grouped together in tightly packed heads with purple bracts between.
FLOWERING PERIOD Apr–Jun.
HABITAT A wide variety of grassy habitats on both calcareous and neutral soils, in woodlands, gardens, roadsides and waste ground.
FREQUENCY Very widespread and common, except at very high altitudes.

Wood Sage

■ *Teucrium scorodonia* Up to 45cm

DESCRIPTION A short, tufted and rather downy perennial plant with Sage-like, but non-aromatic, leaves that have a wrinkled texture and heart-shaped bases. The non-scented yellowish-green flowers, up to 6mm long, grow in opposite pairs on a tall, erect, leafless flower spike, and have prominent stamens and purple-tinted anthers.
FLOWERING PERIOD Jun–Sep.
HABITAT Woodland rides, heaths and cliff tops, usually on acid soils, and often tolerating slightly shady conditions, but also found on limestone pavements and screes.
FREQUENCY Very widespread and common in suitable habitats.

Ground Ivy

■ *Glechoma hederacea* Up to 15cm

DESCRIPTION A low-growing, unpleasant-smelling, softly downy perennial plant, sometimes tinged purple. Long, creeping runners, sometimes reaching 1m in length, send down roots at intervals, and the plant may form complete carpets in places. The bluish-violet flowers are up to 15mm long and grow in loose whorls arising from the leaf axils; they are much favoured by bees in early spring.
FLOWERING PERIOD Mar–Jun.
HABITAT Woodland rides, grassy banks and hedgerows, gardens and waste places, generally preferring fertile soils, but tolerant of sunny or shaded conditions.
FREQUENCY Widespread and often very common, except in the extreme N.

White Dead-nettle
■ *Lamium album* Up to 60cm

DESCRIPTION A low-growing, creeping perennial plant with square stems and very slightly aromatic, downy foliage. The stalked, oval leaves have a heart-shaped base and strongly toothed margin, and greatly resemble those of the **Stinging Nettle** *Urtica dioica*. The white flowers are up to 3cm long; the hooded upper lip is hairy, and the lower lip is toothed and folded back.
FLOWERING PERIOD Mar–Nov.
HABITAT A range of grassy habitats, including woodland margins, roadside verges, scrub patches, disturbed ground and neglected gardens.
FREQUENCY Widespread and very common in most lowland areas, but absent from the far NW and much of Ireland.

Red Dead-nettle
■ *Lamium purpureum* Up to 40cm

DESCRIPTION A low-growing, sprawling annual plant with an overall red tinge to the leaves and, especially, the stems, and a strong scent when crushed. The long-stalked leaves are slightly heart-shaped with rounded teeth on the margins. The dark purple-pink flowers have an overlapping upper lip and a notched, spotted lower lip.
FLOWERING PERIOD Mar–Oct.
HABITAT Wasteland, disturbed ground and bare patches of soil in many locations, from the rural to the most urban, which are rapidly colonised when there is no competition from larger plants.
FREQUENCY Very widespread and common throughout, apart from very acidic or very wet sites.

Common Hemp-nettle
▪ *Galeopsis tetrahit* Up to 50cm

DESCRIPTION An upright, branched, coarsely hairy annual plant with nettle-like leaves in opposite pairs up the stem. The pinkish-purple flowers are up to 2cm long and have darker spots on the lower lip; the calyx is bristly and persists after the flowers have fallen. The flowers are arranged in tightly packed whorls around the stem, with leaves beneath them.
The whole plant has an unpleasant scent.
FLOWERING PERIOD Jul–Sep.
HABITAT Woodland clearings, ditches, hedgerows, disturbed arable ground and waste places, and damp sites such as fens and riverbanks.
FREQUENCY Widespread and locally very common.

Yellow Archangel
▪ *Lamiastrum galeobdolon* Up to 45cm

DESCRIPTION A low-growing, hairy, patch-forming perennial with upright flower stems and rooting runners up to 1m long. The leaves are nettle-like and dark green, and the lower ones have long stalks. The escaped garden variety, ssp. *argenteum*, has silvery blotches on the leaves. The rich golden-yellow flowers have reddish-brown markings on the lower lip and are arranged in tight whorls around the stem.
FLOWERING PERIOD Apr–Jun.
HABITAT Damp woodlands, shady hedgerows, limestone pavements, and sheltered cliffs and banks. Usually considered to be an indicator of ancient woodland.
FREQUENCY Widespread and locally common in the S of the region, and absent from the N. Introduced into scattered locations elsewhere.

Black Horehound
■ *Ballota nigra* Up to 50cm

DESCRIPTION An unpleasant-smelling, hairy perennial plant with oval, toothed, stalked leaves and a square stem; the plant often grows in dense tufts. The pinkish-purple flowers, produced in whorls around the stem, are up to 18mm long and have a concave upper lip; the sepal tube is funnel-shaped and has pointed teeth, and persists after the flowers have fallen.
FLOWERING PERIOD Jun–Sep.
HABITAT Disturbed ground, cultivated areas, wasteland, field margins and hedgerows, usually on nutrient-rich soils and frequently near habitation.
FREQUENCY Locally common in lowland areas in the S and E, but scarce or absent elsewhere.

Betony
■ *Betonica officinalis* Up to 60cm

DESCRIPTION A slightly hairy, long-stemmed perennial plant with a basal rosette of leaves and a few pairs of opposite stem leaves that are ovate with blunt teeth and short stems. The bright reddish-purple flowers are up to 18mm long and are grouped in dense terminal orchid-like heads, making them very conspicuous.
FLOWERING PERIOD Jun–Sep.
HABITAT Open grassy areas, usually in damp conditions, along woodland rides, hedgerows, cliff tops, heathlands and field margins.
FREQUENCY Widespread and locally common in the S and E, but scarce or absent in the N and NW.

Marsh Woundwort
■ *Stachys palustris* Up to 1m

DESCRIPTION A robust, downy and very slightly scented perennial with creeping stems at ground level, and tall, unbranched flowering stems. The leaves are mostly unstalked and very narrow, with heart-shaped bases. The pinkish-purple flowers are arranged in whorls around the stem, and form a slender and attractive flower spike.
FLOWERING PERIOD Jun–Sep.
HABITAT Damp grassland, especially on the margins of rivers and ponds, or in marshes and fens. Also found on old pastures and low-lying areas subject to flooding.
FREQUENCY Widespread and common over a very large area, but more scattered and local in the E.

Hedge Woundwort
■ *Stachys sylvatica* Up to 75cm

DESCRIPTION A roughly hairy perennial plant with a strong, unpleasant smell when bruised. The nettle-like leaves are long-stalked and heart-shaped at the base, and have toothed margins. The flowers are up to 18mm long and reddish purple, with paler spots on the lower lip; they are arranged in whorls around the stem in tall, leafless flower spikes.
FLOWERING PERIOD Jun–Oct.
HABITAT Drier parts of woodlands, hedgerows, roadsides and field margins, and neglected cultivated areas. Mildly acidic or neutral soils are favoured, and it is tolerant of light shade. Can be a troublesome garden weed in some areas.
FREQUENCY Widespread and very common in some lowland areas, but absent from high altitudes.

Water Mint

■ *Mentha aquatica* Up to 50cm

DESCRIPTION A rather variable perennial herb with a stiff, square stem and slightly hairy, oval leaves that have a strong, pleasant smell of mint when crushed. The plant can grow in quite deep water and is sometimes partially submerged. The tiny lilac-pink flowers are rarely more than 4mm long, but they are crowded together in dense terminal heads up to 2cm long and often into further smaller whorls lower down the stem. The flowers are a great attraction to insects. FLOWERING PERIOD Jul–Oct. HABITAT A variety of wet habitats, including ditches, ponds and river margins, plus damp grassland, fens and marshes. FREQUENCY Widespread and often very common in suitable habitats.

Corn Mint

■ *Mentha arvensis* Up to 30cm

DESCRIPTION An upright, but rather weak, hairy perennial herb with a strong, pungent smell of mint. The short-stalked leaves are oval with toothed margins and grow in opposite pairs. Creeping stems send down rootlets at intervals. The tiny flowers are about 4mm long, are grouped in dense whorls at intervals along the stem, and do not form a terminal flower head. FLOWERING PERIOD May–Oct. HABITAT Field margins, grassy banks and hedgerows, woodland rides and the drier areas of marshes and fens; sometimes grows near Water Mint (above), but in drier situations. FREQUENCY Widespread and very common, but becoming scarce in the N and W.

Basil-thyme
■ *Clinopodium acinos* Up to 20cm

DESCRIPTION A low-growing, downy annual herb with both erect and creeping stems. The small leaves are oval with slightly toothed margins and short stalks, and emit no scent if bruised. The small violet flowers, which grow in loose whorls around the stem, are up to 9mm long and have a distinct white patch on the lower lip.
FLOWERING PERIOD May–Aug.
HABITAT Open, sunny areas with low vegetation, usually on dry calcareous soils. Also appears on some newly created roadside verges and embankments.
FREQUENCY Scattered and local across S and E areas, and introduced in Ireland. Absent or very scarce further N and at high altitudes.

Wild Basil
■ *Clinopodium vulgare* Up to 35cm

DESCRIPTION A downy, lightly aromatic perennial herb, usually fairly upright and unbranched, with stalked, ovate leaves that have toothed margins. The pinkish-purple flowers are up to 22mm long and grow in tight whorls around the stem in the axils of the upper leaves; they are supported by bristly purple bracts covered in woolly hairs.
FLOWERING PERIOD Jul–Sep.
HABITAT Dry, usually sunny hedgerows, embankments, woodland and field margins, cliffs, sand-dunes and waste ground. Sometimes troublesome in gardens.
FREQUENCY Widespread and common in much of the S and E, but scarce or absent in the far N and W.

Gipsywort
■ *Lycopus europaeus* Up to 75cm

DESCRIPTION A hairy perennial with a superficially mint-like appearance, but not aromatic. The stems are erect and usually branch from near the base. The long yellowish-green leaves are deeply cut, and are arranged in opposite pairs up the square stem. The tiny white flowers are up to 5mm long and grow in tight whorls in the axils of the upper leaves.
FLOWERING PERIOD Jul–Sep.
HABITAT Damp grassland, pond and river margins, ditches and wet sand-dune slacks; often rapidly colonises newly created wetlands such as gravel pits.
FREQUENCY Widespread and very common in suitable habitats. More scattered in the N and NE.

Wild Marjoram
■ *Origanum vulgare* Up to 50cm

DESCRIPTION A downy perennial herb, usually found growing in dense tufts, with a pleasant scent when bruised. The tough stems usually have a reddish tinge and the oval green leaves grow in opposite pairs. The pinkish-purple flowers are up to 8mm long and are grouped into numerous dense heads on branched flower spikes; they are a great attraction to insects on sunny days.
FLOWERING PERIOD Jul–Sep.
HABITAT Dry, open, sunny grassland areas, most frequently on calcareous soils, and also on railway banks, roadsides, dunes, quarries and as a garden escape.
FREQUENCY Most common on suitable soils in the S and E, becoming scattered and scarce elsewhere.

Wild Thyme

■ *Thymus drucei* Up to 5cm

DESCRIPTION An aromatic, mat-forming perennial herb with a mass of non-flowering stems that send down rootlets, and short, erect flowering stems. The tiny, oval leaves are short-stalked and grow in opposite pairs. The pinkish-purple flowers, with darker purple calyx tubes, are up to 4mm long and grow in dense terminal heads supported on 4-angled stems that bear short hairs on 2 opposite sides.
FLOWERING PERIOD Jun–Sep.
HABITAT Open, sunny sites on free-draining soils and screes, ranging from heathlands and chalk downland to dunes, cliff tops and mountain slopes.
FREQUENCY Widespread and locally very common, but declining where scrub encroaches.

Bittersweet

■ *Solanum dulcamara* Up to 1.5m

DESCRIPTION A very variable, scrambling perennial with a woody base and numerous green shoots that twine through other vegetation; it may be smooth or downy. The long, oval leaves have short stalks and, sometimes, paired lobes at the base. The purple flowers have 5 reflexed lobes and bright yellow projecting anthers; these give rise to clusters of shiny red poisonous berries. The plant is also known as Woody Nightshade.
FLOWERING PERIOD May–Sep.
HABITAT Woodland edges, hedgerows, waste ground, pond and river margins, embankments and scrub, plus coastal shingle and dune slacks.
FREQUENCY Widespread and common, but absent from the far NW.

Great Mullein
▪ *Verbascum thapsus* Up to 2m

DESCRIPTION A striking, tall perennial plant with a strong flower spike, sometimes with candelabra-like branching, and a covering of grey woolly down. The large basal leaves are oval and form a rosette in the 1st year; smaller leaves grow from the flower spike, which emerges in the 2nd year. The yellow flowers are up to 3.5cm across, opening in succession up the stem. Seeds form inside egg-shaped capsules.
FLOWERING PERIOD Jun–Aug.
HABITAT Open grassy sites, such as roadsides, embankments, downland, quarries and waste ground. A rapid coloniser of disturbed ground, and also a garden escape.
FREQUENCY Widespread; commonest in the S, but absent from the far N and high altitudes.

Dark Mullein
▪ *Verbascum nigrum* Up to 1m

DESCRIPTION A slender, upright perennial plant with a distinctive ridged stem. The dark green oval leaves are hairy; the lower leaves grow on long stalks, while the upper ones merge into the stem. The 1–2cm-wide yellow flowers appear to have a dark 'eye' owing to purple hairs on the stamens; they grow in a single tall spike, often all opening at the same time.
FLOWERING PERIOD Jun–Aug.
HABITAT Roadside verges, embankments, arable field margins, disturbed areas on chalk grassland and sand pits.
FREQUENCY Locally common on suitable dry calcareous soils, mainly in the S and E of the region, but appears as a garden escape elsewhere.

Round-leaved Fluellen

■ *Kickxia spuria*
Up to 25cm, but usually creeping

DESCRIPTION An easily overlooked creeping, softly hairy annual plant with rounded or slightly oval leaves growing alternately on short stalks. The flowers have a long, curved spur and are mostly yellow, but have a contrasting purple upper lip; they are usually about 12mm long and are borne singly on slender stalks. Seeds develop inside tiny capsules.
FLOWERING PERIOD Jul–Oct.
HABITAT A 'weed' of cultivated land, usually on dry calcareous lowland soils, but it may also be found on waste ground, tracks and embankments.
FREQUENCY Locally common in the S of the region, and very scarce or absent elsewhere.

Common Figwort

■ *Scrophularia nodosa* Up to 70cm

DESCRIPTION An upright, hairless perennial plant with solid, square stems, and stalked, rather nettle-like lower leaves that have a faint, unpleasant smell when bruised. The curious, almost globular flowers are up to 1cm long and greenish, with a maroon lip and a white border to the sepal lobes; they grow in open, leafless spikes. The seeds are produced inside tiny, rounded capsules.
FLOWERING PERIOD Jun–Sep.
HABITAT Partially shaded grassy areas, including woodland rides, hedgerows, riverbanks, ditches and waste ground.
FREQUENCY Widespread and very common in the S, but becoming more scattered and local in the far N.

Common Toadflax
▪ *Linaria vulgaris* Up to 75cm

DESCRIPTION An attractive upright, hairless perennial plant with greyish-green foliage; the narrow, elongate leaves are borne close to the stem and topped by the flower spike. The yellow 'snapdragon' flowers are up to 2.5cm long and have a curved spur and contrasting orange bulge in the centre. The lower flowers in the spike open first.
FLOWERING PERIOD Jun–Oct.
HABITAT A variety of open grassy habitats, including hedgerows and embankments, roadsides, waste ground and arable field margins.
FREQUENCY Widespread and common in drier lowland areas, becoming more local in the N and at high altitudes.

Ivy-leaved Toadflax
▪ *Cymbalaria muralis* Trailing

DESCRIPTION A hairless perennial plant with purple-tinged trailing stems and small, stalked leaves usually resembling those of Ivy *Hedera helix*. The long-stalked flowers, arising from the leaf axils, are about 10–12mm long and pale purple with a yellow bulge in the centre; hidden from view is a short, curved spur. Fruits are produced on long stalks, which curve back into crevices to 'plant' the seeds.
FLOWERING PERIOD Apr–Nov.
HABITAT Walls, bridges, rocky banks, stabilised shingle and other well-drained, open habitats. An introduced species in Britain and Ireland.
FREQUENCY Very widespread in suitable habitats, including city centres, and absent only from the far N.

Thyme-leaved Speedwell
■ *Veronica serpyllifolia* Up to 20cm

DESCRIPTION A diminutive, sometimes downy perennial plant with creeping stems that send down rootlets, and leafy, upright stems that bear the flowers. The small, oval leaves, resembling those of thyme, grow in opposite pairs. The flowers are borne on short stalks arising from the leaf axils, and are rarely more than 7mm across; they can be pale blue or white. Seeds are produced inside flattened capsules.
FLOWERING PERIOD Apr–Oct.
HABITAT A variety of grassy habitats, ranging from garden lawns to woodland rides, heaths, rocky ledges, mountain slopes and damp flushes.
FREQUENCY Widespread and common throughout, including city centres.

Germander Speedwell
■ *Veronica chamaedrys* Up to 20cm

DESCRIPTION A delicate but eye-catching hairy perennial plant with creeping stems that send down rootlets at intervals. The opposite, short-stalked leaves are oval and have toothed margins. The flowers are borne on slender stalks that have 2 opposite lines of hairs running along them, and are usually produced prolifically in good conditions. Each 4-petalled flower is up to 12mm across and bright blue with a white centre. The fruits are a flattened heart shape.
FLOWERING PERIOD Apr–Jun.
HABITAT A wide range of open grassy places, including woodland rides, hedgerows, embankments, cultivated ground, rocky ledges and screes, reaching quite high altitudes.
FREQUENCY Very widespread and common in most areas.

Heath Speedwell
■ *Veronica officinalis* Up to 10cm

DESCRIPTION A low-growing, downy perennial plant that forms mats of creeping stems, from which the narrow flowering shoots arise. The stems are hairy all around and roots grow down from the joints. The leaves are hairy, oval, slightly toothed and unstalked, and grow in opposite pairs. The pale lilac flowers are up to 9mm across and open from the base upwards in slender spikes.
FLOWERING PERIOD May–Aug.
HABITAT Grassy areas, including woodland rides and clearings, grassy banks, heathlands, anthills and dune slacks.
FREQUENCY Very widespread and common wherever there are well-drained soils, reaching high altitudes.

Blue Water-speedwell
■ *Veronica anagallis-aquatica* Up to 25cm

DESCRIPTION A hairless perennial plant with rooting, creeping stems that are sometimes submerged, and upright flowering shoots. The leaves are narrow and long, sometimes measuring as much as 12cm, with short, pointed teeth. The flowers are borne in dense, paired flower spikes that emerge on long stalks from the leaf axils. Individual flowers are 4-lobed and pale blue. The very similar **Pink Water-speedwell** V. *catenata* has pink flowers and short flower stalks.
FLOWERING PERIOD Jun–Aug.
HABITAT Damp sites such as pond and stream margins, canals, neglected Watercress beds, ditches and woodland rides.
FREQUENCY Locally common in the S and E, but scarce or absent elsewhere.

Common Field Speedwell ■ *Veronica persica* Creeping

DESCRIPTION A creeping, rather hairy and much-branched annual plant that has reddish stems and alternate or almost paired oval, toothed, pale green leaves. The 4-petalled flowers are up to 8mm across and pale blue, but the lowest petal is almost white. The flowers are supported singly on slender stalks arising from the leaf axils. The seeds are produced in broad, flattened capsules.
FLOWERING PERIOD Jan–Dec.
HABITAT Arable fields, waste ground, gardens, tracks and roadsides, wherever there is bare or disturbed soil.
FREQUENCY Accidentally introduced to W Europe and the UK from the Caucasus in early 19th century, and now very widespread and common, especially in the S and E.

Monkeyflower
■ *Erythranthe guttata* Up to 50cm

DESCRIPTION An upright and very showy perennial plant with downy upperparts and hairless, leafy runners that creep along the ground. The leaves are broadly oval and toothed, and the lower ones are stalked. The bright yellow flowers are up to 4.5cm long and borne on short stalks; there is a bulge in the 3-lobed lower lip, which is marked with red spots, while the 2-lobed upper lip is unspotted.
FLOWERING PERIOD Jun–Sep.
HABITAT Marshy ground in sunny situations, such as riverbanks, ditches, fens and mildly acid bogs.
FREQUENCY Introduced from North America; now widely naturalised and becoming common in some areas.

Foxglove

▪ *Digitalis purpurea* Up to 1.5m

DESCRIPTION A tall, robust and very attractive biennial or short-lived perennial plant that produces tapering flower spikes. In the 1st year, the downy, oval, slightly wrinkled leaves form a basal rosette; the flower spike emerges from this in the 2nd year. The pinkish-purple, or occasionally white, flowers are up to 5cm long and tubular, with a scattering of darker spots in the throat. Seeds are produced inside rounded capsules. FLOWERING PERIOD Jun–Sep. HABITAT Woodlands, hedgerows, moorlands, sea cliffs, rocky hillsides and mountain slopes, on acid soils. FREQUENCY Widespread, and may be abundant in suitable habitats.

Eyebright

▪ *Euphrasia officinalis* agg. Up to 25cm

DESCRIPTION A low-growing, branched, hairy annual plant with oval, deeply toothed leaves that are usually a dark, glossy green. Many similar species and confusing hybrids occur, and there is considerable variation. The flowers, which grow in leafy spikes, are up to 10mm long (depending on the species) and the lower lobe is deeply divided. Most flowers are white with purple veins and a yellow throat. Eyebrights are semi-parasitic on grasses. FLOWERING PERIOD May–Sep. HABITAT A wide range of open, undisturbed grassy habitats, including dune slacks, heaths and moors, chalk downland and mountain slopes. FREQUENCY Widespread and locally common, but declining in range owing to habitat loss.

Red Bartsia

■ *Odontites vernus* Up to 40cm

DESCRIPTION A downy, rather
straggling annual plant with a reddish
tinge to the stems and leaves. The
leaves are narrow and unstalked,
growing in opposite pairs. The pinkish-
purple flowers are up to 10mm long and
2-lipped; the lower lip is further divided
into 3 lobes. The flowers are arranged
in a slightly drooping, 1-sided spike
with numerous leaf-like bracts below
the flowers. A semi-parasite on the
roots of grasses.
FLOWERING PERIOD Jul–Sep.
HABITAT Grassy areas, including field
margins, roadsides, dune slacks, waste
ground, tracks and open woodland rides.
FREQUENCY Widespread and common
throughout, although more local in the
far N.

Common Cow-wheat

■ *Melampyrum pratense* Up to 35cm

DESCRIPTION A low-growing,
usually hairless annual plant that may
be branched or straggle through grasses.
The leaves are long and narrow, and
grow in opposite pairs. The yellow
flowers grow in pairs in the axils of the
leaves; they are up to 18mm long, are
held horizontally and look flattened
from side to side, and the mouth of the
flower is almost closed. A semi-parasite
on the roots of other plants.
FLOWERING PERIOD May–Sep.
HABITAT Grassy woodland rides,
heaths and moors, often on acid soils,
but may grow on calcareous soils in
lowland areas.
FREQUENCY Widespread and locally
common, especially in the N and W.

Yellow Rattle
▪ *Rhinanthus minor* Up to 45cm

DESCRIPTION An upright, usually hairless annual plant with a stiff, 4-angled stem that is often marked with purple blotches. The opposite pairs of leaves are a narrow, oblong shape with toothed margins. The flattened yellow flowers are about 20mm long and have a slightly hooded upper lip. The seeds are produced in the inflated papery calyx, which makes a rattling sound when shaken. A semi-parasite on the roots of other plants.
FLOWERING PERIOD May–Sep.
HABITAT Undisturbed grassland such as chalk downland, traditional hay meadows and dry dune slacks.
FREQUENCY Widespread, and may be abundant in some good sites.

Lousewort
▪ *Pedicularis sylvatica* Up to 20cm

DESCRIPTION A low-growing, much-branched, hairless perennial plant with almost feathery leaves deeply divided into numerous toothed leaflets. The pale pink flowers are about 25mm long and 2-lipped, the upper lip being notched, and have an inflated purplish calyx at the base. They are slightly flattened in appearance. Up to 10 flowers grow together in a terminal head, and the seeds develop inside an inflated capsule. A semi-parasite on the roots of other plants.
FLOWERING PERIOD Apr–Jul.
HABITAT Damp acid grassland, wet flushes, drier parts of bogs and marshes, reaching quite high altitudes.
FREQUENCY Widespread and locally common in suitable habitats, becoming more scarce in the drier SE.

Marsh Lousewort

■ *Pedicularis palustris* Up to 60cm

DESCRIPTION A hairless, upright perennial plant with a single branching stem and leaves that are deeply divided into numerous feathery lobes. The pinkish-purple flowers are up to 25mm long and 2-lipped, the upper lip having 4 teeth; they grow in open leafy spikes, with leaf-like bracts below the upper flowers. The seeds are produced inside inflated capsules. A semi-parasite on the roots of other plants.
FLOWERING PERIOD May–Sep.
HABITAT Wet grassland, marshes, bogs and other wetlands such as fens; not confined to acid soil conditions. Does not occur at very high altitudes.
FREQUENCY Widespread and locally common, but less so in the S and E of the region.

Toothwort

■ *Lathraea squamaria* Up to 25cm

DESCRIPTION A curious low-growing perennial plant that has just a few pale scales on the stem and no green leaves, and is entirely parasitic on the roots of woody shrubs and trees; it is most commonly found below Hazel. The pinkish-white flowers are up to 18mm long and tubular, growing in a dense 1-sided spike with a fanciful resemblance to a row of teeth, hence its common name. Seeds develop inside a capsule within the dead flower.
FLOWERING PERIOD Apr–May.
HABITAT Native deciduous woodlands and hedgerows, invariably on calcareous soils and always in lowland areas.
FREQUENCY Widespread but only locally common in suitable habitats, and absent from N Scotland, W Ireland and N Scandinavia.

Common Broomrape
■ *Orobanche minor* Up to 40cm

DESCRIPTION An unusual upright, unbranched perennial plant that is entirely parasitic on the roots of other plants, most commonly clovers and trefoils, and consequently has no green leaves. The single stem has a purplish tinge and supports the numerous slightly downcurved, tubular flowers. Each flower is up to 18mm long and pinkish yellow with darker purple veins and a purple stigma. Many similar species occur, each associated with a specific host plant.
FLOWERING PERIOD Jun–Sep.
HABITAT Open grassy areas, including chalk downland, dune slacks, roadsides and embankments.
FREQUENCY Locally common in lowland regions, but scarce or absent elsewhere.

Moschatel
■ *Adoxa moschatellina* Up to 10cm

DESCRIPTION A small, hairless perennial plant, sometimes so abundant that it carpets the ground. The small 4- or 5-petalled greenish-yellow flowers are up to 8mm across and arranged in heads of 5 on a slender stalk; 4 of the flowers face outwards, like clock faces, and the 5th faces upwards. The leaves are pale green and rather fleshy; the lower ones are twice 3-lobed, while the upper leaves are 3-lobed and have longer stalks.
FLOWERING PERIOD Apr–May.
HABITAT Damp, shaded woodlands, usually on heavy soils; also sometimes on banks along shady lanes and in mountain gullies.
FREQUENCY Locally common in the S and E, scarce elsewhere, and absent from Ireland and N Scandinavia.

Common Valerian
▪ *Valeriana officinalis* Up to 1.5m

DESCRIPTION A tall, mostly hairless
perennial plant, usually unbranched,
with toothed, pinnately divided
leaves that end in a single leaflet.
The small, 5-petalled flowers are
usually pale pink, sometimes almost
white, and 5mm across, and are
grouped in a dense terminal head
on a leafless flower stalk. The seeds
have a feathery pappus to aid wind
dispersal.
FLOWERING PERIOD Jun–Aug.
HABITAT A variety of grassy
habitats, including woodland rides,
waysides, river and stream margins,
mountain grasslands and rough chalk
grassland.
FREQUENCY Widespread and locally
common.

Common Butterwort
▪ *Pinguicula vulgaris* Up to 15cm

DESCRIPTION A stickily hairy
insectivorous perennial plant. The
basal rosette of oval yellowish-green
leaves have slightly inrolled margins
and act as a trap for small insects.
Leafless flower stalks arise from the
rosette, bearing the single 25mm-
long flowers. The flowers are violet-
coloured with a white throat, and
shaped like a funnel with a long spur.
FLOWERING PERIOD May–Aug.
HABITAT Bare, damp peaty exposures
on moors and mountains, wet flushes,
shallow ditches and sphagnum bogs;
also on mossy patches in fens.
FREQUENCY Widespread and locally
common in the N and W, now very
scarce and declining in the S and E.

Greater Plantain
■ *Plantago major* Up to 20cm

DESCRIPTION A low-growing, sometimes downy perennial plant with a basal rosette of broadly oval leaves on long leaf stalks, which usually grows flat on the ground. The minute yellowish-white flowers, no more than 3mm across, are compacted into a dense head on a tall flower stalk, opening in succession up the stem. The anthers are purple at first and turn yellow later. FLOWERING PERIOD May–Aug. HABITAT Disturbed and cultivated ground, lawns, paths and tracks, sports fields and parks, and many other grassy areas with low vegetation. Very tolerant of trampling, mowing and disturbance. FREQUENCY Widespread and very common throughout entire region.

Honeysuckle
■ *Lonicera periclymenum*
Climbing, up to 5m

DESCRIPTION A deciduous woody climber with numerous twisting stems that twine over trees, shrubs and each other in a clockwise direction. The leaves are ovate, in pairs, and may have short stalks. The terminal heads of long, tubular flowers, reaching 5cm at times, have a 4-lobed upper lip and entire lower lip, and are white or creamy yellow; they are very sweetly scented, especially at night. The fruits are clusters of shiny red berries. FLOWERING PERIOD Jun–Aug. HABITAT Woodlands, scrub, hedgerows and, sometimes, scrambling on rocky slopes. FREQUENCY Widespread and common throughout, but scarce in Scandinavia.

Twinflower

■ *Linnaea borealis* Up to 7cm

DESCRIPTION A delicate, creeping, downy evergreen perennial plant, which sometimes forms mats of stems and leaves on the ground in sheltered woodlands. The rounded leaves grow in pairs on thread-like stems. The attractive pink bell-shaped flowers grow in pairs on a slender, upright flower stalk; each flower is about 8mm long and nodding.
FLOWERING PERIOD Jun–Aug.
HABITAT Scottish pinewoods, and occasionally in more open heathland habitats.
FREQUENCY Restricted to ancient and recently planted pinewoods in NE Scotland, and occasionally found in N England, where it may have been accidentally introduced; widespread and common in Scandinavia.

Field Scabious

■ *Knautia arvensis* Up to 75cm

DESCRIPTION An upright, roughly hairy, branched perennial plant with an overwintering basal rosette of lobed leaves, and more variable stem leaves. Flattish heads, up to 4cm across, of bluish-violet flowers with pink anthers are supported on long stalks; each flower is made up of 4 unequal petal lobes, and the outer flowers are larger than the inner ones.
FLOWERING PERIOD Jun–Oct.
HABITAT Dry grassland in sunny situations, roadsides, hedgerows and chalk downland.
FREQUENCY Widespread and locally common on suitable sites, but scarce in N Scotland, W Ireland and N Scandinavia.

Devil's-bit Scabious

■ *Succisa pratensis* Up to 75cm

DESCRIPTION An upright, branching perennial plant that may either be smooth or slightly hairy. The lower leaves are spoon-shaped, and the few stem leaves are narrow and sparsely toothed. Globular flower heads up to 2.5cm across grow on tall, slender flower stalks; each tiny, 4-petalled flower is bluish violet with projecting anthers that resemble tiny mallets. FLOWERING PERIOD Jun–Oct. HABITAT Damp grassland, woodland rides and clearings, marshes, heathlands, and damp chalk and limestone, sometimes at high altitudes. FREQUENCY Widespread and sometimes locally common.

Wild Teasel

■ *Dipsacus fullonum* Up to 2m

DESCRIPTION A very tall, upright biennial plant with branched stems that are hairless but spiny. In the 1st year, the leaves form a basal rosette, but this withers away in the 2nd spring. The basal leaves are very long and covered with swollen spines, and the stem leaves clasp the stem, trapping pools of water in which insects drown. The flowers grow in tightly packed ovoid heads protected by spiny bracts. Seeds form in the dried flower head. FLOWERING PERIOD Jul–Aug. HABITAT Rough grassland, roadsides, waste ground, spoil tips, embankments and gravel pits. FREQUENCY Widespread and common in the S, but local or rare in the N.

Sheep's-bit

■ *Jasione montana* Up to 40cm

DESCRIPTION A low-growing, sometimes creeping, downy biennial plant with a basal rosette of oblong, short-stalked, wavy-edged leaves, and stalkless stem leaves. The sky-blue flowers grow in globular heads up to 35mm across; each flower is only about 5mm across and tubular, splitting nearly to the base when open, but without projecting anthers. FLOWERING PERIOD May–Sep. HABITAT Dry grassland in sunny situations, including coastal rocks and cliffs, heaths and dunes, embankments and hedgerows on acid soils, and avoiding chalk. FREQUENCY Widespread, but common only near the sea and in the W. Absent from large areas of the N and E.

Harebell

■ *Campanula rotundifolia* Up to 40cm

DESCRIPTION A slender, hairless perennial plant with thin, wiry stems and narrow stem leaves, making it difficult to find when not in flower. The basal leaves are rounded and long-stalked, but often wither away. The attractive nodding, bell-shaped blue flowers are up to 15mm long and have triangular teeth, and grow in loose spikes. This is the 'Bluebell' of Scotland. FLOWERING PERIOD Jul–Sep. HABITAT Open grassy areas, including heaths and chalk downland, dune slacks, moorland and mountain slopes, and cliffs and screes, often reaching very high altitudes. FREQUENCY Widespread and common, but quite scarce in SW England, Ireland and Brittany.

Nettle-leaved Bellflower
■ *Campanula trachelium* Up to 75cm

DESCRIPTION A tall, roughly hairy perennial plant with sharp-angled stems and short-stalked stem leaves that bear a close resemblance to those of **Stinging Nettles** *Urtica dioica*; the basal leaves have heart-shaped bases and longer stalks. The attractive bluish-violet flowers are up to 4cm long and have wide-open triangular lobes. They are produced in tall, leafy spikes.
FLOWERING PERIOD Jul–Aug.
HABITAT Hedgerows, scrub, woodland rides and shady banks, usually on heavy calcareous soils.
FREQUENCY Locally common in the S and E of England, absent or very scarce elsewhere.

Daisy
■ *Bellis perennis*
Up to 10cm

DESCRIPTION A very familiar downy perennial plant with rosettes of spoon-shaped leaves that grow flat to the ground and are tolerant of heavy trampling. The often prolifically produced flower heads are supported on leafless, hairy stalks. The outer florets are white, tinged with pink on the underside, and the inner florets forming the disc are yellow.
FLOWERING PERIOD Mar–Oct.
HABITAT Lawns, sports fields, pastures, roadside verges, dune slacks and upland grasslands.
FREQUENCY Widespread and common in almost all areas.

Oxeye Daisy

■ *Leucanthemum vulgare* Up to 70cm

DESCRIPTION A downy or sometimes hairy perennial plant with 8cm-long spoon-shaped, blunt-toothed basal leaves and narrow, stalkless stem leaves. The flowers are supported on long, sometimes branched stems, and can be up to 6cm across. The outer, or ray, florets are all white, and the inner, or disc, florets are yellow, with no scales between them.
FLOWERING PERIOD Jul–Sep.
HABITAT Dry grasslands in sunny situations, including downlands, dune slacks, roadsides, cliffs and embankments, favouring neutral or calcareous soils.
FREQUENCY A widespread and common species in most areas, and becoming very familiar on motorway verges.

Scentless Mayweed

■ *Tripleurospermum inodorum* Up to 60cm

DESCRIPTION A scentless, hairless, straggling annual plant with deeply divided, almost feathery leaves. The daisy-like flower heads, which are up to 4cm across, are produced singly on long stalks; they have white ray florets and yellow disc florets, with brown-edged sepal-like bracts below.
FLOWERING PERIOD Apr–Oct.
HABITAT Disturbed and cultivated ground, tracks, roadsides and field margins, often forming extensive patches.
FREQUENCY Very common and widespread in most areas, but scarce in N Scotland and W Ireland.

Sea Mayweed
■ *Tripleurospermum maritimum*
Up to 60cm

DESCRIPTION A perennial plant, very similar to Scentless Mayweed (p. 119) but more branched and spreading, and often mat-forming on coastal shingle. The lower stems sometimes have a purplish tinge. The leaflets are shorter than those of Scentless Mayweed, and also slightly succulent. The daisy-like flowers are up to 4cm across and solitary, borne on long stalks.
FLOWERING PERIOD Apr–Oct.
HABITAT Coastal shingle and sand, cliffs, sea walls and embankments, and disturbed ground near the sea.
FREQUENCY Very common and widespread in most areas, but scarce in N Scotland, W Ireland and N Scandinavia.

Pineapple Mayweed
■ *Matricaria discoidea* Up to 30cm

DESCRIPTION A much-branched, hairless, bright green perennial plant that smells strongly of pineapple when crushed. The leaves are finely divided, giving them a feathery appearance. The solitary yellow flowering heads grow on short stalks and are up to 8mm across and conical, consisting only of tightly packed disc florets; there are no ray florets.
FLOWERING PERIOD Jun–Aug.
HABITAT Disturbed fertile ground on farms or in gardens, often on well-trodden paths or in gateways, and also on roadsides and urban sites.
FREQUENCY Widespread and common throughout Britain, having been introduced here in the late 19th century. Scattered and more local in Norway.

Sea Aster

■ *Tripolium pannonicum* Up to 75cm

DESCRIPTION An upright, hairless, fairly short-lived seaside perennial plant with long, spear-shaped, slightly fleshy leaves that have a prominent mid-rib. The flowers grow in loose clusters on leafy shoots, each flower head consisting of bluish-lilac ray florets and yellow disc florets, and measuring up to 2cm across. Beware confusion with garden escape Michaelmas daisies, which are less likely in maritime conditions.
FLOWERING PERIOD Jul–Sep.
HABITAT The sheltered upper reaches of saltmarshes, along the shores of estuaries, in brackish ditches, and on sea cliffs and sea walls.
FREQUENCY Locally common along suitable stretches of coastline around the whole of the region.

Blue Fleabane

■ *Erigeron acris* Up to 30cm

DESCRIPTION A slender, upright, hairy annual or biennial plant with a rather stiff stem that may be tinged with red. The leaves at the base of the stem are spoon-shaped and stalked, but the stem leaves are narrow and unstalked. The flower heads are held up on long stalks and consist of an outer circle of bluish-purple ray florets that almost conceal the yellow disc florets.
FLOWERING PERIOD Jun–Aug.
HABITAT Dry, sunny, permanent grassland sites, including stable sand-dunes and shingle, old walls and banks, and even industrial waste sites.
FREQUENCY Widespread and locally common in England and Wales, but scarce or absent in Ireland and Scotland; very common elsewhere in the region.

Common Cudweed
■ *Filago germanica* Up to 25cm

DESCRIPTION A small but upright, densely hairy annual with a greyish-white woolly appearance. The narrow leaves have wavy margins and grow close to the stem, which usually branches towards the top. The flowers form tight, rounded clusters measuring about 12mm across and consisting of 20–35 tiny flower heads, each of which is composed of tiny yellow florets surrounded by yellowish bracts.
FLOWERING PERIOD Jul–Aug.
HABITAT Dry grassy sites, including heathlands, arable field margins, tracks, sand-dunes and waste ground, generally preferring sandy soils.
FREQUENCY Locally common in the S, but declining, and scarce or largely absent elsewhere.

Marsh Cudweed ■ *Gnaphalium uliginosum* Up to 20cm

DESCRIPTION A multi-branched, creeping or erect annual plant with a grey-green woolly appearance. The leaves are narrow and woolly on both surfaces, and the uppermost

leaves surround the flower heads, often overtopping them. The flowers grow in stalkless heads about 4mm long, and are made up of yellow disc florets surrounded by papery brown bracts.
FLOWERING PERIOD Jul–Oct.
HABITAT Open grassy areas, often on sites subjected to trampling, and usually on damp or seasonally waterlogged soils, including clay and sand. May also turn up in parks and gardens.
FREQUENCY Widespread and common.

Ploughman's-spikenard
■ *Inula conyzae* Up to 1m

DESCRIPTION An upright, downy biennial or short-lived perennial plant, often with red-tinged stems. Its basal leaves are oval and may be mistaken for those of Foxgloves (p. 108), but the stem leaves are long and narrow. The ovoid flower heads grow in clusters, each head measuring about 12mm across and consisting of yellow disc florets and purplish-green bracts.
FLOWERING PERIOD Jul–Sep.
HABITAT Dry, sunny grassland sites on calcareous soils, including dunes, chalk downland, quarries, roadsides and embankments.
FREQUENCY Locally common in the S and E, but more scattered elsewhere on suitable sites, and absent from the N of the region.

Golden Samphire
■ *Limbarda crithmoides*
Up to 75cm

DESCRIPTION An eye-catching upright, tufted perennial with abundant heads of showy flowers and narrow but fleshy, hairless, bright green leaves. The flower heads are up to 30mm across and grow in loose terminal clusters. They are made up of an outer ring of spreading yellow ray florets and an inner section of orange-yellow disc florets.
FLOWERING PERIOD Jul–Sep.
HABITAT Saltmarshes, stabilised coastal shingle and calcareous sea cliffs.
FREQUENCY Widespread, but only locally common around the coasts of S England and Wales, SE Ireland and N France.

Common Fleabane
▪ *Pulicaria dysenterica* Up to 50cm

DESCRIPTION A spreading perennial plant with numerous upright, branched, rather woolly stems. The oblong stem leaves have heart-shaped bases that clasp the stem and are hairy on both surfaces. The flower heads grow in loose terminal clusters, each head measuring about 30mm across, and are composed of spreading yellow ray florets and deeper yellow disc florets.
FLOWERING PERIOD Jul–Sep.
HABITAT Damp meadows, roadside verges, fens, wet dune slacks, pond and stream margins.
FREQUENCY Widespread and common in the S and W, scattered and local elsewhere and largely absent from the N of the region.

Trifid Bur-marigold
▪ *Bidens tripartita* Up to 60cm

DESCRIPTION An almost hairless annual plant with reddish branched stems. The stem leaves grow on winged leaf stalks and are deeply 3-lobed, with toothed margins. The flowers grow in upright heads up to 30mm across and are made up of tiny yellow disc florets surrounded by green bracts. The very similar **Nodding Bur-marigold** B. *cernua* has hairy stems and drooping flower heads.
FLOWERING PERIOD Jul–Oct.
HABITAT Damp grassy areas and shallow water at the margins of ponds, streams and rivers, and in sluggish drainage ditches.
FREQUENCY Locally common in the S and E, but scarce or absent further N.

Goldenrod

▪ *Solidago virgaurea* Up to 75cm

DESCRIPTION A rather variable, usually
unbranched perennial plant that may
be slightly downy. The basal leaves are
long-stalked and spoon-shaped, but the
stem leaves are narrow and unstalked,
and sometimes have toothed margins.
The yellow flowers are up to 10mm
across and are made up of both ray and
disc florets, and they are arranged in
narrow, branched spikes.
FLOWERING PERIOD Jun–Sep.
HABITAT Woodlands on acid soils,
lanes and hedgerows, heathlands,
dune slacks, cliffs, rocky hillsides
and mountain ledges.
FREQUENCY Very widespread and
locally common, but scarce in E England.

Yarrow

▪ *Achillea millefolium* Up to 50cm

DESCRIPTION A strongly aromatic,
upright, quite downy perennial plant
with creeping stems and erect, furrowed
flowering shoots. The leaves are dark
green and finely divided, with a feathery
appearance. The flowers grow in a dense,
flat-topped cluster of many small heads;
each head, measuring about 6mm across,
comprises yellowish disc florets and
pinkish-white ray florets. Even when
it is not in flower, Yarrow's presence is
obvious from the carpets of aromatic
foliage it forms in open grasslands.
FLOWERING PERIOD Jun–Nov.
HABITAT A wide variety of grassy
habitats, including meadows, roadsides,
cliff tops, dunes, upland slopes, mountain
ledges and waste ground.
FREQUENCY Widespread and very
common throughout.

Sneezewort

■ *Achillea ptarmica* Up to 60cm

DESCRIPTION An upright, usually branched perennial plant with furrowed, slightly hairy stems. The leaves are narrow and unstalked, with finely toothed margins. The flower heads are up to 2cm across and comprise greenish-yellow disc florets and large white ray florets; they are grouped together in flat-topped clusters. A very similar double-flowered garden escape variety may sometimes be encountered near habitation.
FLOWERING PERIOD Jul–Sep.
HABITAT Damp grassland, usually on acid soils, in woodland rides, heathlands, moorlands and mountain slopes.
FREQUENCY Widespread and locally common in suitable habitats.

Corn Marigold

■ *Glebionis segetum* Up to 50cm

DESCRIPTION A very attractive, conspicuous hairless annual plant with greyish-green aromatic foliage. The slightly fleshy leaves are narrow and deeply toothed or lobed. The flower heads are solitary and up to 45mm across, and are made up of orange-yellow disc florets and large yellow ray florets. The similar but more robust and darker orange **Pot Marigold** *Calendula officinalis* may appear in the wild as a garden escape.
FLOWERING PERIOD Jun–Oct.
HABITAT Disturbed ground and neglected arable land, field margins and sandy roadsides, preferring well-drained acid soils.
FREQUENCY Once a familiar 'weed of cultivation' but now declining; although widespread, it is only very locally common.

Tansy

■ *Tanacetum vulgare* Up to 1m

DESCRIPTION A strongly aromatic, upright perennial plant with long, pinnately divided yellowish-green leaves, each leaflet having toothed margins. The flowers form compact umbel-like clusters of up to 70 flower heads, each one button-like, about 10mm across and consisting of yellow disc florets only.
FLOWERING PERIOD Jul–Oct.
HABITAT Roadside verges, hedgerows, embankments, waste ground, riverbanks and arable margins.
FREQUENCY Widespread and very common in the S and E, but scarce in the N and NW.

Feverfew

■ *Tanacetum parthenium* Up to 50cm

DESCRIPTION A strongly aromatic, much-branched, quite downy perennial plant with yellowish-green foliage. The leaves are pinnately divided, the lower ones growing on short stalks and the upper ones stalkless. The daisy-like flowers grow in loose clusters, each flower measuring up to 2cm across and comprising white ray florets and yellow disc florets.
FLOWERING PERIOD Jul–Aug.
HABITAT Originally introduced as a garden plant and now widely naturalised in grassy places, waste ground, old walls, tips and embankments.
FREQUENCY Widespread across most of the region, but absent from the far N and at high altitudes.

Hemp-agrimony
■ *Eupatorium cannabinum* Up to 1.5m

DESCRIPTION A very tall, robust, downy perennial plant, often with red-tinged stems. The leaves are divided into 3, or sometimes 5, toothed lobes and grow in opposite pairs up the stem. The pinkish-lilac flowers form dense, frothy terminal clusters, acting as a great attraction to insects, especially butterflies and moths. Each tiny flower head is about 5mm across and made up of about 5 florets.
FLOWERING PERIOD Jul–Sep.
HABITAT Damp grassland, pond and stream margins, ditches, chalk scrub and roadsides.
FREQUENCY Widespread and locally common in the S and E, but confined mainly to the coastal region in Scotland, and absent from the far N.

Butterbur
■ *Petasites hybridus* Up to 50cm

DESCRIPTION A low-growing, patch-forming perennial plant with enormous leaves resembling those of rhubarb; by the end of the season they may be 1m across, with a long stalk, heart-shaped base and grey-downy underside. The pinkish-red flowers are produced before the leaves, with male and female flowers on separate plants. Male flowers are 7–12mm across and female flowers about 3–6mm across. They are clustered in thick spikes, appearing long before the leaves.
FLOWERING PERIOD Mar–May.
HABITAT Damp grassy areas, often beside rivers and streams, and cliffs and gullies where springs emerge.
FREQUENCY Widespread in the S and E, but scarce or absent in the N.

Mugwort

■ *Artemisia vulgaris* Up to 1.25m

DESCRIPTION An upright, slightly aromatic, downy perennial plant with grey-green foliage and ribbed reddish stems. The leaves are pinnately divided and green on the upper surface, but covered with silvery down on the underside. The flowers grow in small reddish heads about 3mm across, arranged in tall, branched, slightly leafy spikes.
FLOWERING PERIOD Jul–Sep.
HABITAT Grassy places, including roadsides, hedgerows, waste ground, embankments and urban sites.
FREQUENCY Widespread and locally common in lowland areas, but very scarce in the N and in upland areas.

Common Ragwort

■ *Jacobaea vulgaris* Up to 1m

DESCRIPTION A hairless, and poisonous, biennial plant with a stiff stem, branched above the middle, and pinnately divided leaves terminating in a blunt lobe. The yellow flower heads are about 15–25mm across and made up of long ray florets and darker yellow disc florets; they are grouped together in flat-topped clusters. The foliage is the favourite food plant of the black and orange caterpillars of the Cinnabar Moth *Tyria jacobaeae*.
FLOWERING PERIOD Jun–Nov.
HABITAT Dry grassy places, thriving in heavily grazed areas, and also on roadsides, dune slacks and waste ground, both urban and rural.
FREQUENCY Widespread and very common throughout, and possibly increasing.

Groundsel
■ *Senecio vulgaris* Up to 40cm

DESCRIPTION A branching annual plant, often with downy foliage. The leaves are pinnately lobed, the lower ones having stalks and the upper ones clasping the stem, which is usually red-tinged and slightly ribbed. The flowers are yellow and cylindrical, rarely more than 10mm long, and composed of tiny disc florets only and supported by black-tipped bracts. The seeds have a hairy 'parachute'.
FLOWERING PERIOD Jan–Dec.
HABITAT A well-known weed of gardens, arable land, waste ground and industrial sites. Also occurs on dunes and shingle.
FREQUENCY Widespread and very common, apart from in the far N and at high altitudes.

Coltsfoot
■ *Tussilago farfara* Up to 15cm

DESCRIPTION A creeping perennial plant that spreads by means of runners, and with numerous upright, leafless flowering shoots that appear in spring before the leaves. The 10–20cm-wide leaves emerge after the flowers and are rounded, with heart-shaped bases. The scaly, rather woolly purplish flower stalks support the yellow heads, which are up to 35mm across and made up of both disc and ray florets. The seeds form a rounded head of white feathery 'parachutes'.
FLOWERING PERIOD Feb–Apr.
HABITAT Disturbed ground, including landslips, spoil tips, roadsides, shingle banks, dune slacks, wet flushes, cliffs and mountain ledges.
FREQUENCY Widespread and very common in places, reaching quite high altitudes.

Greater Burdock
■ *Arctium lappa* Up to 1m

DESCRIPTION A large, much branched, downy biennial plant with broad, heart-shaped leaves supported on solid stems. The rounded flower heads are up to 4cm across and made up of numerous tiny purple florets, these just visible inside the cup of hooked, spiny bracts. This then forms a prickly brown bur, which hooks into passing animals or humans to aid seed dispersal.
FLOWERING PERIOD Jul–Sep.
HABITAT Hedgerows, woodland rides and clearings, arable margins, roadsides and waste ground.
FREQUENCY Locally common in England and Wales, but scarce or absent elsewhere.

Musk Thistle
■ *Carduus nutans* Up to 1m

DESCRIPTION An upright and rather elegant biennial thistle. The cottony stems have spiny wings on the lower sections, while the upper sections of the flower stalks are usually free of spines. The solitary nodding flower heads are about 4cm across and made up of numerous reddish-purple florets and spiny purple-tinged bracts.
FLOWERING PERIOD Jun–Aug.
HABITAT Rough grasslands, roadsides, waste ground and open scrubby sites, preferring dry calcareous soils.
FREQUENCY Locally common in the S and E, but scarce or absent elsewhere, and never at high altitudes.

Welted Thistle
■ *Carduus crispus* Up to 1.3m

DESCRIPTION A very tall, branched thistle with spiny, winged stems but with a smooth section immediately below the flower heads. The leaves are long and pinnately lobed, with many sharp spines, the upper ones being smaller and unstalked. The flower heads are roughly egg-shaped, and about 2–3cm long with reddish-purple florets and woolly green bracts.
FLOWERING PERIOD Jun–Aug.
HABITAT Roadsides, hedgerows, waste ground, field margins and rough grassland, generally in lowland areas and on fertile soils.
FREQUENCY Locally common in the S and E, but scarce or largely absent in Scotland and Ireland, and absent from high altitudes.

Carline Thistle
■ *Carlina vulgaris* Up to 60cm

DESCRIPTION An upright, thistle-like biennial plant with numerous stiff spines. The leaves are pinnately lobed, with spiny points and a hairy upper surface. The usually solitary flower heads are golden brown, up to 3cm across and surrounded by straw-coloured bracts. The dead flower heads persist long into the winter, as do the spiny leaves.
FLOWERING PERIOD Jul–Sep.
HABITAT Close-cropped dry grassland on calcareous soils, including chalk downland, cliffs and dunes.
FREQUENCY Locally common, mainly in the S and E, and largely absent from Scotland and N Scandinavia.

Spear Thistle

■ *Cirsium vulgare* Up to 1m

DESCRIPTION An upright biennial thistle, with downy stems that have spiny wings between the leaves. The leaves themselves are very long, spiny and pinnately lobed, the upper ones clasping the stem. The purple flower heads are up to 3cm across and globular, with spiny bracts at the base; the heads may be solitary or grow in small clusters. The seeds have long, feathery 'parachutes'. FLOWERING PERIOD Jul–Sep. HABITAT A range of grassland habitats, including pastures, arable margins, roadsides, waste ground and derelict urban sites. Usually more common on fertile soils. FREQUENCY Widespread and very common, reaching quite high altitudes.

Meadow Thistle

■ *Cirsium dissectum* Up to 75cm

DESCRIPTION A medium-height, downy perennial thistle that sends out runners and has unwinged, rather slender, ridged flowering stems. The lower leaves are long and undivided, with a green upper surface and downy white underside. The flowers grow in solitary heads, up to 25mm across, with purple florets and darker bracts. The seeds have white feathery 'parachutes' to aid dispersal. FLOWERING PERIOD Jun–Aug. HABITAT Wet meadows, damp hollows on heathlands, old pastures and wet flushes in acid or calcareous conditions. FREQUENCY Locally common in the SW of the region, widespread in Ireland, and absent from the N and E.

Creeping Thistle
■ *Cirsium arvense* Up to 1m

DESCRIPTION A rather variable perennial thistle with a creeping rootstock and unwinged stems. The leaves are pinnately lobed and spiny, with a shiny upper surface; the topmost leaves clasp the stem. The fragrant purple flower heads are up to 2.5cm across and are made up of numerous pinkish-lilac florets and darker bracts.
FLOWERING PERIOD Jul–Sep.
HABITAT A wide range of grassland habitats, including overgrazed meadows, roadsides and embankments, arable margins, waste ground and urban sites.
FREQUENCY Widespread and very common throughout, apart from at very high altitudes. May form dense patches in places, especially on disturbed soils.

Marsh Thistle
■ *Cirsium palustre* Up to 2m

DESCRIPTION A very tall, slender biennial thistle with stems that are spiny-winged and leafy to the top. The stem leaves are long, pinnately lobed and very spiny with a hairy surface, and they clasp the stem. The flower heads are about 15mm across and grow in terminal clusters on short stalks; the florets are dark purple, but pure white forms sometimes occur. The bracts below the florets are purple-tipped and fairly soft.
FLOWERING PERIOD Jul–Sep.
HABITAT Wet pastures, damp woodland rides, fens and marshes, wet flushes on mountain and moorland slopes.
FREQUENCY Widespread and common throughout.

Common Knapweed

■ *Centaurea nigra* Up to 1m

DESCRIPTION An attractive downy, or sometimes hairy, perennial plant with stiff, grooved stems that are slightly swollen below the flower heads. The usually solitary brush-like flower heads are up to 4cm across, with reddish-purple florets and a solid, swollen base covered in brown bracts. Much appreciated by insects. Also known as Black Knapweed.
FLOWERING PERIOD Jun–Sep.
HABITAT Rough grasslands, old meadows, downland, roadsides, cliff tops and waste ground, often covering large areas.
FREQUENCY Widespread and common throughout, apart from at high altitudes.

Greater Knapweed

■ *Centaurea scabiosa* Up to 1m

DESCRIPTION An attractive perennial plant with stiff, downy stems that are swollen towards the base and grooved, and oblong, pinnately divided leaves. The solitary flower heads are up to 5cm across, supported on long flower stalks, and have many deep reddish-purple florets, the outer ones spreading and divided like ray florets. The base of the flower head is hard and swollen, and is covered with scaly brown bracts.
FLOWERING PERIOD Jun–Sep.
HABITAT Rough grassland, chalk downland, scrub, hedgerows, roadside verges, traditional meadows, cliff tops and dunes, almost always on calcareous soils.
FREQUENCY Locally common in the S and E, and scattered in a few other coastal locations. Not found at high altitudes.

Goat's-beard

■ *Tragopogon pratensis* Up to 70cm

DESCRIPTION An upright, slender annual or short-lived perennial plant. The long, pointed, smooth leaves have a conspicuous white mid-rib and sheath the stem at their bases. If cut, the stem and leaves exude a milky sap. The Dandelion-like flower heads are up to 4cm across and made up of yellow florets and long bracts. They open fully only on sunny mornings and then close up again by midday. The seeds are produced in a large white 'clock' up to 10cm across, each seed having a large 'parachute'.
FLOWERING PERIOD Jun–Jul.
HABITAT Rough grassland, roadsides, sand-dunes, waste ground, arable field margins, tracks and paths.
FREQUENCY Widespread and common in the S and E, scattered and more local in the N and NW.

Perennial-sowthistle

■ *Sonchus arvensis* Up to 1.5m

DESCRIPTION A very tall perennial plant with erect flowering stems. The very long, pinnately divided leaves have spiny teeth on their margins and clasp the stem with rounded basal lobes. The eye-catching bright yellow flowers are up to 5cm across and have bracts below them covered in sticky hairs. They are arranged in loose, umbel-like clusters. The seeds form a large 'clock'.
FLOWERING PERIOD Jul–Sep.
HABITAT Damp grassy places, disturbed ground, wasteland, arable margins, and the upper parts of saltmarshes and sheltered strand lines.
FREQUENCY Widespread and very common in the S, more coastal and scattered in the N.

Common Dandelion
■ *Taraxacum officinale* agg. Up to 35cm

DESCRIPTION An extremely variable perennial plant, but a very familiar one. The leaves form a basal rosette, sometimes flat on the ground and sometimes growing up through grass. The leaves are long and deeply toothed along the margins. The hollow flower stalks exude a milky latex. The bright yellow flowers are made up of numerous ray florets, and the outer ones may be red on the underside. The seeds form the familiar 'dandelion clock'. Many very similar and confusing species occur.
FLOWERING PERIOD Mar–Oct.
HABITAT A wide range of grassy habitats, from coastal sands to mountain ledges, but especially common in gardens, parks, roadsides and agricultural areas.
FREQUENCY Widespread and very common in all areas, sometimes creating a blaze of yellow flowers in spring.

Mouse-ear Hawkweed
■ *Pilosella officinarum* Up to 25cm

DESCRIPTION A low-growing perennial plant whose leaves and stem are covered in white hairs. The leaves are very hairy on the underside, looking white rather than green and feeling very soft. Slender, sparsely leafy runners are produced from the basal rosette. The solitary lemon-yellow flowers, which are up to 3cm across, have red-tinged outer florets, and bracts covered with dark hairs.
FLOWERING PERIOD May–Oct.
HABITAT Short grassland, heaths, dunes, embankments, chalk downland, quarries, rock outcrops, cliffs and mountain slopes.
FREQUENCY Widespread and common throughout.

Hawkweed Oxtongue
■ *Picris hieracioides* Up to 70cm

DESCRIPTION A tall perennial plant, one of many very similar and rather confusing species with yellow flowers resembling those of the Common Dandelion (p. 137). The stem is hairy and often reddish near the base, and branches frequently. The long leaves have margins that are wavy rather than toothed. The yellow flowers are up to 25mm across, and 1 row of the bracts below them is downcurved. FLOWERING PERIOD Jul–Oct. HABITAT Dry grassland, usually on chalk or limestone, roadsides, embankments and quarries, but not on sites with heavy grazing. FREQUENCY Widespread in the S and E of the region. Largely absent elsewhere.

Autumn Hawkbit
■ *Scorzoneroides autumnalis* Up to 25cm

DESCRIPTION A rather variable, hairless, or occasionally slightly hairy, perennial plant. The long, deeply lobed leaves resemble those of the Common Dandelion (p. 137) and form a basal rosette. The yellow flowers are in heads up to 35mm across and are usually solitary, although some stems may be branched. Below the flower head, which tapers towards its base, the stem bears many scale-like bracts. FLOWERING PERIOD Jun–Oct. HABITAT A wide range of grassy habitats, usually preferring acid soils, from sea-level to high mountain ledges, and including dunes, cliff tops, roadsides, heaths, moorland and upland lake margins. FREQUENCY Common and widespread throughout, including urban centres.

Common Catsear

■ *Hypochaeris radicata* Up to 50cm

DESCRIPTION A tufted perennial plant that has numerous slender, hairless stems, some of them branched, with tiny, scale-like bracts, these fancifully resembling a cat's ears. The leaves are deeply lobed, roughly hairy and grow in a basal rosette. The yellow flowers form heads up to 40mm across, with short, bristly, purple-tipped bracts below them on a slightly swollen stalk.
FLOWERING PERIOD Jun–Sep.
HABITAT A wide range of dry grassy habitats, usually on slightly acid soils, ranging from sand-dunes and cliff tops, through lawns and roadsides, to heathlands and waste ground.
FREQUENCY Widespread and common throughout, although absent from N Scandinavia and scarce at high altitudes and on waterlogged sites.

Bog Asphodel

■ *Narthecium ossifragum* Up to 30cm

DESCRIPTION A tufted, hairless perennial plant with very attractive spikes of yellow flowers. The narrow leaves, resembling miniature iris leaves, form a flat fan-like tuft at the base of the plant. The star-like yellow flowers are up to 15mm across and have orange anthers. The seeds are produced inside orange capsules, and the whole plant turns orange and persists through the winter. Slightly poisonous to grazing animals.
FLOWERING PERIOD Jun–Aug.
HABITAT Boggy ground in heaths and moorlands, especially in valley and blanket bogs, and often reaching quite high altitudes; always in open, sunny situations.
FREQUENCY Widespread in suitable habitats in N and W regions, but very scarce and local in the SE.

Bluebell
▪ *Hyacinthoides non-scripta* Up to 50cm

DESCRIPTION A hairless, bulbous perennial plant with very long, glossy green leaves arising from the base and hollow flower stalks bearing the fragrant, 6-lobed, bell-shaped blue flowers in a drooping, 1-sided spike. Very occasionally, white or pale pink flowers occur. The similar but non-native **Spanish Bluebell** *H. hispanica* has much paler flowers growing all around the stalk in an upright spike.

FLOWERING PERIOD Apr–Jun.

HABITAT Woodlands, especially ancient broadleaf woodland and Hazel coppice, hedgerows, upland grassland under Bracken, cliff tops and parkland. Often forms complete carpets in stable habitats.

FREQUENCY Widespread and common over a large area of the W, but becoming scarce in the E and absent from very high altitudes and some N islands.

Ramsons ▪ *Allium ursinum* Up to 35cm

DESCRIPTION A hairless, bulbous perennial plant with a distinct smell of garlic. The leaves arise from the base and are long and oval, with a long, often twisted leaf stalk. The white flowers are up to 20mm across and bell-shaped, growing in large, rounded heads on leafless flower stalks. This species can be so abundant that it often forms complete carpets on woodland floors or along riverbanks.

FLOWERING PERIOD Apr–May.

HABITAT Damp woodland on heavy soils, riverbanks, shady hedgerows, rock crevices, scree slopes and sheltered sea cliffs. Usually grows best in shaded situations.

FREQUENCY Widespread and sometimes locally common, but more scarce and scattered in N and E areas.

Lily-of-the-valley

■ *Convallaria majalis* Up to 20cm

DESCRIPTION A creeping perennial plant with long, branching rhizomes and hairless, paired, long-stalked oval leaves. The sweetly scented, pure white flowers are bell-shaped, long-stalked and nodding, and grow in a drooping, 1-sided spike. These are followed after pollination by red berries.
FLOWERING PERIOD May–Jun.
HABITAT Dry, open woodlands, most frequently on thin but calcareous soils, especially Ash woods on limestone. Needs light shade to flower prolifically.
FREQUENCY Widely scattered as a native plant except in the far N; also occurs as a frequent garden escape in many areas.

Common Solomon's-seal

■ *Polygonatum multiflorum* Up to 60cm

DESCRIPTION A creeping, hairless perennial plant with long, arched, leafy stems and alternate oval leaves growing directly from the stem. The greenish-white bell-shaped flowers have a pinched 'waist' and hang down in 2s or 3s on slender stalks that arise from the leaf axils. The fruits are in the form of bluish-black berries. A slightly more robust form, an escape from gardens, is sometimes seen in woodlands.
FLOWERING PERIOD May–Jun.
HABITAT Woods and copses on calcareous or neutral soils, and occasionally in hedgerows and churchyards, and on roadsides. Very susceptible to grazing by deer.
FREQUENCY Locally common in the S and E, but may appear as a garden escape in other areas.

Herb Paris
■ *Paris quadrifolia* Up to 35cm

DESCRIPTION A very distinctive, hairless perennial plant with 4 broad, oval leaves that have prominent veins and are arranged in a flat whorl at the top of the stem. The single flower, which is slightly raised above the leaves, has 8 yellow stamens surrounded by very slender yellowish-green petals and sepals. The fruit is a shiny black berry.
FLOWERING PERIOD May–Jun.
HABITAT Damp, partly shaded woodland clearings and rides, usually on calcareous soils, and almost always in ancient woodland sites; also in grykes in limestone pavements.
FREQUENCY Locally common in England and Wales on suitable soil types, but more widespread and common in the E of the region.

Butcher's Broom
■ *Ruscus aculeatus* Up to 1m

DESCRIPTION A curious evergreen shrub with numerous stiff branches and spiny, leaf-like, flattened stems called cladodes, which bear the tiny stalkless flowers; the true leaves are minute and scale-like. The tiny flowers are only about 3mm across and grow directly on the cladodes, with male and female flowers on separate plants. The female flowers give rise to large red berries.
FLOWERING PERIOD Jan–Mar.
HABITAT Dry woodlands, hedgerows, coastal cliffs and scrub; frequently occurs as a garden escape in churchyards and on roadsides near habitation.
FREQUENCY Locally common only in the S of the region, but more scattered elsewhere, where it is likely to be a garden escape.

Wild Daffodil

■ *Narcissus pseudonarcissus* Up to 50cm

DESCRIPTION A hairless, bulbous perennial plant easily recognised but sometimes confused with similar garden escapes. The leaves are grey-green and narrow, and all arise from the base of the plant. The pleasantly scented flowers are about 5cm across and are protected by a papery bract before opening; they have 6 yellow outer segments and a darker yellow trumpet in the centre.
FLOWERING PERIOD Mar–Apr.
HABITAT Damp, open woodlands, copses and hedgerows, and valley grasslands, often forming extensive carpets, but declining if shaded out by other plants.
FREQUENCY Widespread and locally common in the S and W, but introduced or a garden escape in other areas, where it soon becomes naturalised.

Snowdrop

■ *Galanthus nivalis* Up to 25cm

DESCRIPTION A familiar and much-loved bulbous perennial of early spring, with long, narrow, hairless grey-green leaves arising from the base. The solitary nodding white flowers have 3 pure white outer segments and 3 white inner segments, each with a green patch. Seeds are produced in capsules, but the plants mainly spread by division of the bulbs, leading to the formation of carpets of flowers.
FLOWERING PERIOD Feb–Mar.
HABITAT Shady woodlands, hedgerows, churchyards, parks and large gardens.
FREQUENCY Widespread and, in places, very common, but almost certainly introduced to Britain and much of Europe, and present as an escape in most areas. Very scarce in N Scotland, Ireland and Scandinavia.

Yellow Iris
■ *Iris pseudacorus* Up to 1m

DESCRIPTION A very robust, showy, hairless perennial plant with rows of sword-shaped leaves, and thick, partly submerged rhizomes from which the smooth, occasionally branched stems arise. The attractive yellow flowers are up to 10cm across and have faint purplish veins. These are followed by swollen seed pods. Sometimes known as the Yellow Flag.
FLOWERING PERIOD May–Aug.
HABITAT A variety of wet habitats, including pond and river margins, ditches and canals, wet meadows and marshes, dune slacks and slightly brackish coastal marshes.
FREQUENCY Very widespread and, in places, abundant. On many offshore islands, but not found at high altitudes.

Lords-and-ladies
■ *Arum maculatum* Up to 50cm

DESCRIPTION A hairless perennial plant with large, arrow-shaped, long-stalked leaves that may have many dark blotches on them and reach a length of over 20cm. The curious flowers consist of a purple-edged pale green enveloping spathe, which is pinched in at the bottom to protect the base of the purple central spike, or spadix, of male and female flowers. After pollination, a spike of bright orange-red berries develops. Also known as Cuckoo-pint.
FLOWERING PERIOD Apr–May.
HABITAT Damp woodlands, copses, hedgerows, cliffs and gullies, generally on fertile soils and in partial shade. Also creeps into parks and gardens.
FREQUENCY Widespread and very common in the S, but probably introduced to Scotland and the Isle of Man. Absent from most of Scandinavia and many upland regions.

Lady's Slipper

■ *Cypripedium calceolus* Up to 50cm

DESCRIPTION An unmistakable and very beautiful orchid with a large, colourful flower up to 5cm across. The inflated lower lip is bright yellow, and there are 3 tassel-like maroon outer segments. 1 or 2 flowers grow on a single flower stalk. The large, hairless leaves resemble those of Lily-of-the-valley (p. 141), and when not in flower this orchid is easily overlooked.
FLOWERING PERIOD May–Jun.
HABITAT Open limestone woodlands and copses, and occasionally more open grassland sites.
FREQUENCY Scattered and local in suitable protected sites across the N and E of the region; extremely rare in Britain and well guarded at its very few locations in N England.

Fly Orchid

■ *Ophrys insectifera* Up to 40cm

DESCRIPTION A slender, rather inconspicuous orchid, easily overlooked in tall grasses, even when in flower. The oval, glossy leaves form a basal rosette and some grow up the stem. The flowers bear a passing resemblance to bees or flies, having a velvety lower lip in the shape of an insect, 2 upper petals looking like antennae and 3 green sepals. Glossy patches help create the illusion of an insect's body.
FLOWERING PERIOD May–Jun.
HABITAT Dry grassy areas, open woodlands and scrub, chalk pits, quarries and embankments, almost always preferring lime-rich and well-drained sites.
FREQUENCY Widely scattered and only locally common across most of the region.

Bee Orchid ■ *Ophrys apifera* Up to 30cm

DESCRIPTION An attractive and very distinctive orchid, with a spike of up to 10 flowers vaguely resembling bumblebees. The 3 sepals are pink, and the upper petals are slender and green, but the lower lip is greatly inflated and furry, with variable markings. There is a long, tooth-like yellowish lobe hidden from view below the lip. The leaves are oval, green and hairless; there are 2 stem leaves, with the remaining leaves forming a basal rosette.
FLOWERING PERIOD Jun–Jul.
HABITAT Dry, sunny grassland sites, usually on chalky soils but sometimes on industrial waste ground. Often found on roadsides, and in quarries and gravel pits.
FREQUENCY Widespread and locally common in S and E England, but more scattered elsewhere and absent from the N of the region. Often colonises newly created grassy banks.

Early Purple Orchid
■ *Orchis mascula* Up to 50cm

DESCRIPTION A very attractive orchid, which produces tall spikes of purplish-crimson flowers, although very pale pink forms occasionally occur. Each flower has a 3-lobed lower lip with reflexed side lobes, and a long, upward-curving spur. The shiny leaves, which form a basal rosette, usually have bold, dark blotches.
FLOWERING PERIOD Apr–Jun.
HABITAT Damp woodlands and copses, usually on neutral soils, and also on more open grassy sites, hedgerows, cliff ledges, limestone pavements, roadsides and embankments.
FREQUENCY Widespread and locally common across the whole region, but declining in some woodlands through pressure from deer grazing.

Green-winged Orchid

■ *Anacamptis morio* Up to 40cm

DESCRIPTION A small orchid with spikes of purple flowers and glossy, unspotted leaves that form a basal rosette and sheath the base of the stem. The flowers show great variation in colour, ranging from pure white through shades of pink to pinkish purple. The upper petals have distinct dark veins and often a greenish tinge; they are reflexed to form the 'wings'.
FLOWERING PERIOD Apr–Jun.
HABITAT Old, permanent pastures and hay meadows, dune slacks, churchyards and coastal grasslands.
FREQUENCY Widely scattered in the S and E, and very scarce in the N; generally declining owing to the ploughing of grassland sites.

Burnt Orchid

■ *Neotinea ustulata* Up to 20cm

DESCRIPTION A delightful small orchid that often produces dense tufts of flower spikes, each of which is tightly packed with small maroon buds that become paler and eventually white when fully open. The lower lip is 3-lobed and marked with dark spots. The flowers have an unusual vanilla scent, and although rather small are very conspicuous in short grass. The leaves are unspotted and rather a dull green, and form a basal rosette.
FLOWERING PERIOD May–Jul.
HABITAT Open, sunny sites on dry chalk and limestone grassland, especially where there are plenty of grazing animals to reduce scrub.
FREQUENCY Locally common in the S and E of the region, but declining owing to the loss of good-quality grassland sites.

Bird's-nest Orchid
■ *Neottia nidus-avis* Up to 35cm

DESCRIPTION A very unusual orchid with no green coloration or true leaves. It obtains its nutrition as a parasite of fungi living below the soil, and has a tangle of roots resembling a bird's nest. The flowers grow in a tall, cylindrical spike and are brownish; they have a hood, but no spur and a 2-lobed lip. When fully open, they release a delicate scent of honey to attract pollinating insects.
FLOWERING PERIOD May–Jul.
HABITAT Densely shaded woodlands, often of Beech, where there is a chalky soil and a rich humus layer. Sometimes also found in Hazel and Ash coppices.
FREQUENCY Widely scattered but never very common, and absent from the far N.

Common Twayblade
■ *Neottia ovata* Up to 60cm

DESCRIPTION An easily overlooked orchid, with a basal pair of broad, 16cm-long, oval leaves and a slender spike of rather small greenish-yellow flowers. The leaves appear long before the flowers, and the lax flower spike then grows up from the centre. The flowers have a deeply forked yellowish lower lip and a hooded appearance due to the converging sepals and petals.
FLOWERING PERIOD May–Jul.
HABITAT Open woodlands, copses, scrub, embankments, quarries, limestone pavements, dune slacks and other grassy sites, on a wide range of soil types.
FREQUENCY Widespread and locally common, especially in the S, and sometimes forming large colonies.

Pyramidal Orchid

■ *Anacamptis pyramidalis* Up to 35cm

DESCRIPTION An attractive, delicately scented orchid with dense conical or rounded spikes of pink flowers. The individual flowers have a distinctly 3-lobed lower lip and a long, curved spur. The hairless leaves are greyish green and rather narrow, usually growing upright and partially sheathing the stem.
FLOWERING PERIOD Jun–Aug.
HABITAT Open, sunny sites on well-drained calcareous soils, including chalk grassland, dunes, cliff tops, roadside verges and embankments, limestone pavements and quarries.
FREQUENCY Locally common in the S and E, more scattered elsewhere and on only a few coastal sites in the N.

Fragrant Orchid

■ *Gymnadenia conopsea* Up to 45cm

DESCRIPTION A robust, relatively tall orchid with a dense cylindrical spike of very fragrant pink flowers. Each small flower has a slightly lobed lower lip and a very long, curved spur. The leaves are long and narrow, and partially surround the stem. 3 very similar species of fragrant orchid are known to occur in different habitats.
FLOWERING PERIOD Jun–Aug.
HABITAT Chalk grassland, marshy grassland, heaths, fens and man-made sites such as roadside verges, quarries and waste ground.
FREQUENCY Very widespread and locally common in suitable habitats, becoming more frequent on well-managed roadside embankments.

Frog Orchid
■ *Coeloglossum viride* Up to 25cm

DESCRIPTION A small, inconspicuous orchid, hard to find among grasses even when in full flower. Broad, oval leaves form a basal rosette, and narrower leaves partially sheath the stem. The greenish-yellow flowers have a hooded appearance due to the converging outer petals and sepals, and the longer lower lip is notched at the tip; they grow in a short spike of up to 20 flowers, elongating when in seed.
FLOWERING PERIOD Jun–Aug.
HABITAT A range of open grassy habitats, including chalk downland, sand-dunes, chalk pits, embankments, limestone pavements, and upland cliffs and ledges.
FREQUENCY Very widespread and locally common in the S on dry grassland and in the far N on upland pastures.

Common Spotted-orchid
■ *Dactylorhiza fuchsii* Up to 60cm

DESCRIPTION A very variable orchid, but always with long, narrow, boldly spotted leaves, some at ground level and others growing up the stem. The flowers form a dense cylindrical spike, and can range in colour from pure white through shades of pink to a deeper purple. The lower lip has 3 even-sized lobes and is marked with darker streaks and spots.
FLOWERING PERIOD May–Aug.
HABITAT A wide range of grassy habitats, including woodland rides and clearings, roadsides, sand-dunes, scrub, marshes, quarries, waste ground, churchyards and large gardens.
FREQUENCY The commonest orchid of the region, and very widespread, although more scarce in the far N and extreme SW.

Heath Spotted-orchid

■ *Dactylorhiza maculata* Up to 50cm

DESCRIPTION Resembles the Common Spotted-orchid (p. 150) but is smaller with a more conical flower head that bears fewer flowers. The flowers are usually very pale pink with darker markings, although dark pinkish-purple forms do occur, and the lower lip has a tiny central lobe and a slightly frilled margin. The very narrow leaves have rounded spots.
FLOWERING PERIOD May–Aug.
HABITAT Damp, mostly acidic soils in old pastures, and especially on heaths and moors, usually in full sun. Also occurs on mountain ledges and upland grasslands.
FREQUENCY Widespread and locally common over most of region, but more scarce in the SE.

Early Marsh-orchid

■ *Dactylorhiza incarnata* Up to 60cm

DESCRIPTION A robust orchid with long, unspotted yellowish-green leaves, sometimes with a hooded tip. The flowers grow in a compact spike and are usually a flesh-pink colour, but creamy-white or reddish-purple forms occur in some subspecies. The flower lip has 3 lobes and is marked with darker spots and lines.
FLOWERING PERIOD May–Jun.
HABITAT Wet grassland sites, especially on calcareous soils, including marshes, fens, dune slacks and, occasionally, heaths. May often occur with other species of orchid.
FREQUENCY Widely scattered in suitable habitats, but rarely very common. Declining in some areas owing to drainage and ploughing.

Southern Marsh-orchid
■ *Dactylorhiza praetermissa* Up to 70cm

DESCRIPTION A robust orchid with a stout flower spike of pinkish-purple flowers. The long leaves are a glossy dark green and very infrequently have dark rings; the upper ones clasp the stem. The flowers have a broad, 3-lobed lip with fine, dark markings and a short, blunt spur. The rather similar **Northern Marsh-orchid** *D. purpurella* is usually shorter with darker purple flowers and a diamond-shaped lip.
FLOWERING PERIOD May–Jun.
HABITAT Damp grassland, marshes, fens and dune slacks, and also on seasonally flooded wasteland and abandoned industrial sites.
FREQUENCY Widespread and common in the S; replaced by Northern Marsh-orchid in similar habitats further N.

Greater Butterfly-orchid
■ *Platanthera chlorantha* Up to 50cm

DESCRIPTION A tall, slender, elegant orchid with a graceful spike of delicately scented greenish-white flowers. There are 2 large, oval, unspotted leaves at the base of the stem, and a few very small stem leaves. The flowers have a long, narrow lip and a very long, curved spur; the yellowish pollen sacs form an inverted 'V'. The very similar **Lesser Butterfly-orchid** *P. bifolia* has parallel pollen sacs.
FLOWERING PERIOD Jun–Jul.
HABITAT Usually found on well-drained calcareous soils, including chalk downland, rough pasture, old meadows, scrub, hedgerows and embankments.
FREQUENCY Widespread but only locally common. The Lesser Butterfly-orchid has similar range, but occurs on more acid soils and in more upland regions, and may grow in large colonies.

White Helleborine

■ *Cephalanthera damasonium* Up to 50cm

DESCRIPTION An attractive orchid with
tall, leafless spikes of pure creamy-white
flowers that rarely open fully. The leaves
are green and unspotted, and broad at the
base of the plant but smaller up the stem.
The flowers are up to 20mm long and
bell-shaped, held upwards on the stem.
They are usually self-pollinated as they
do not open up completely, but the yellow
anthers can just be glimpsed inside. The
Sword-leaved Helleborine C. *longifolia*
is very similar but the flowers open up
more and grow in a leafy spike.
FLOWERING PERIOD May–Jul.
HABITAT Beech woods, especially where
there is little ground cover, but also in
scrub on chalky soils and, occasionally,
on roadsides.
FREQUENCY Locally common only in
the S and E; occasionally colonises new
sites in Beech woods.

Marsh Helleborine

■ *Epipactis palustris* Up to 50cm

DESCRIPTION An attractive medium-
sized orchid with broadly oval, unspotted
leaves arranged spirally around the stem.
The open spikes of up to 15 flowers are
predominantly pale in colour. The outer
sepals are usually brownish purple with
green markings and hairs on the outer
surface, and the inner petals are much paler
with red streaks. A very pale colour form
sometimes occurs alongside the darker ones.
FLOWERING PERIOD Jul–Aug.
HABITAT Very damp pastures, wet flushes,
fens and marshes; also very frequent in
wet dune slacks and on coastal landslips.
FREQUENCY Locally common in the S
and E, scattered and very rare elsewhere.

Broad-leaved Helleborine
■ *Epipactis helleborine* Up to 80cm

DESCRIPTION A tall, clump-forming orchid with long, rather downy stems and dense flower spikes. The leaves are broadly oval with pronounced veins. The flowers are drooping and greenish red, with a heart-shaped purple-tinged lip that has a recurved tip and purple-tinged petals above. The flowers have an unpleasant scent and are pollinated by wasps and night-flying insects.
FLOWERING PERIOD Jul–Sep.
HABITAT Shady woodlands, scrub, hedgerows, roadsides, churchyards, parks and neglected gardens, on acid and alkaline soils.
FREQUENCY Very widespread and locally quite common in the S of the region, even in urban sites. Scarce or absent in the far N.

Autumn Lady's-tresses
■ *Spiranthes spiralis* Up to 15cm

DESCRIPTION An easily overlooked, small orchid, which has tightly packed, spirally twisted spikes of pure white flowers. The oval leaves form a basal rosette but usually wither away before the flowers appear. The tiny, delicately scented flowers are about 7mm long and have downy petals and sepals; the open centre of the flower is green.
FLOWERING PERIOD Aug–Sep.
HABITAT A range of short, dry grassland habitats, ranging from coastal dunes and cliff tops to chalk grassland, lawns and roadside verges.
FREQUENCY Locally common in the S and E, often vegetating for many years without flowering until grazing or mowing stops, when it may appear in hundreds.

White Water-lily
■ *Nymphaea alba* Aquatic

DESCRIPTION A very conspicuous aquatic plant during the summer, when the large, circular leaves are visible; they die back completely in winter. The most attractive white flowers are protected inside pointed green buds and open fully only in bright sunshine; they are held just above the surface of the water on stout stalks, and have a delicate scent. Can grow in depths of up to 3m.
FLOWERING PERIOD Jun–Aug.
HABITAT Still or slow-flowing freshwater habitats such as canals, ponds, fenland drainage ditches and shallow lakes. Tolerant of both acid and alkaline conditions.
FREQUENCY Widespread and locally common, but subject to damage by boats on busy waterways.

Yellow Water-lily
■ *Nuphar lutea* Aquatic

DESCRIPTION An aquatic plant with both floating and submerged leaves that are up to 40cm long and leathery, with overlapping basal lobes and prominent veins. The rounded yellow flowers are up to 6cm across and are held above the water surface on long stalks; they have a curious alcohol scent to attract pollinating insects. The seeds develop inside flagon-shaped pods.
FLOWERING PERIOD Jun–Sep.
HABITAT Generally slow-moving or still water in rivers, canals, lakes and ponds; tolerant of partial shade and some disturbance by boats.
FREQUENCY Widespread and locally common, but not present in the far N and probably introduced to the SW.

REFERENCES AND FURTHER READING

Fitter, A. (1978). *An Atlas of the Wild Flowers of Britain and Northern Europe*. Collins.

Mossberg, B. and Stenberg, L. (2003). *Den Nya Nordiska Floran*. Wahlström and Widstrand.

Preston, C.D., Pearman, D.A. and Dines, T.D. (2002). *New Atlas of the British and Irish Flora*. Oxford University Press.

Rose, F. (2006). *The Wild Flower Key*. 2nd edn. Warne.

Stace, C. A. (2019). *New Flora of the British Isles*. 4th edn. Cambridge University Press.

Sterry, P. (2006). *Complete British Wild Flowers*. Collins.

Streeter, D. (2009). *Collins Flower Guide*. Collins.

USEFUL ADDRESSES AND CONTACTS

Botanical Society of the British Isles (BSBI)
Botany Department
The Natural History Museum
Cromwell Road
London
SW7 5BD
www.bsbi.org.uk

Field Studies Council (FSC)
Preston Montford
Shrewsbury
Shropshire
SY4 1HW
www.field-studies-council.org

Plantlife
14 Rollestone Street
Salisbury
Wiltshire
SP1 1DX
www.plantlife.org.uk

The Wildlife Trusts
The Kiln
Waterside Mather Road
Newark
Nottinghamshire
NG24 1WT
www.wildlifetrusts.org